Forms of Corrosion RECOGNITION AND PREVENTION

Dale McIntrye, Editor

Volume 2

NACE HANDBOOK 1

NACE International

Library of Congress Catalog Card Number: 96-070184
ISBN: 1-57590-026-2

NACE International
1440 South Creek, Houston, TX 77084 (713) 492-0535 www.nace.org

Contents

Preface

When I undertook the task of editing a second volume of C.P. Dillon's *Forms of Corrosion: Recognition and Prevention*, my brief charge from the NACE Publications Committee was to expand the work, not alter it. The steady and continuing demand for the first edition even now, 14 years after its original publication, is ample tribute to the classic relevance of the case histories selected by the original editor.

Accordingly, this second volume preserves the basic structure of the first and its classic descriptions of the basic forms of corrosion. This new volume includes more than 180 additional case histories of corrosion failures.

These additional cases have been selected from a broad variety of industries and equipment beyond the utilities and chemical processing plants which provided the bulk of the cases for the first edition. Included are cases from oil and gas production, petroleum refining, cross-country pipelines, marine and land transport, aircraft, and electronics. As a reminder that corrosion affects activities outside of heavy industry, some intriguing cases of corrosion problems in the preservation of historical structures and artifacts are presented.

This second volume expands on some of the "classic" corrosion and cracking problems endemic to certain industrial equipment. Examples include external stress-corrosion cracking of buried cross-country pipelines in alkaline soils, boiler feedwater deaerator cracking, and hydrogen-induced cracking of pipelines and pressure vessels. Summaries of some notable catastrophic failures which have occurred since the first edition's publication are included as worst case examples.

The analytical tools used to study any given corrosion failure vary widely, depending not only on the phenomenon involved but also on the risks and consequences of recurrence. A 40-year-old cast iron soil pipe might rate less than a minute's visual examination before replacement in kind. By contrast, even the simplest general corrosion failures in lethal service provoke in-depth analysis, including failures whose cause is apparent to the unaided eye. At the other extreme are novel processes or subtle environmental effects whose investigation may benefit from sophisticated analytical techniques.

Because the depth of analysis will vary with the objective of the investigation, this new volume includes cases using a wide variety of analytical tools. The results of optical metallography, scanning electron microscopy, energy-dispersive x-ray spectrometry, Aüger spectroscopy, and electrochemical techniques, such as potentiodynamic voltammetry and scratch testing are all presented to illustrate their utility in certain aspects of corrosion failure investigations.

Corrosion prevention techniques fall broadly into five categories:

1. Selection of resistant materials,
2. Environmental control,
3. Barrier coatings,
4. Electrochemical potential control, and
5. Design to eliminate localized aggressive environments.

Many industrial items give economical service lives without complete corrosion prevention. Indeed, absolute suppression of corrosion may be so expensive as to render a structure or process non-competitive. Consequently, many widely used corrosion remedies reduce, but do not eliminate, corrosion. Decisions about the level of protection required for any particular item must be made on a case-by-case basis, weighing the consequences of failure, desired service life, and the financial requirements of management with regard to rate of return on investment.

For this reason, there is seldom only one "correct" solution to a corrosion problem. The remedies described in these case histories must not be taken as absolutes. They are the solutions deemed appropriate by the responsible corrosion engineer in each particular case, given the risks, rewards, time constraints and prior history that governed the particular incident. Other alternatives might prove as good or better in similar circumstances at a different time.

The best tool for analyzing corrosion failures is the mind of a well-trained, experienced corrosion engineer. The corrosion engineer stands at the intersection between science and technology, blending the practical and theoretical aspects of corrosion science to produce workable solutions to realistic problems.

This second volume of *Forms of Corrosion: Recognition and Prevention* is dedicated to these corrosion engineers in the hope that it will aid them in their work.

Dale R. McIntyre
Saudi Aramco
Dhahran, Saudi Arabia

CHAPTER 1

General Corrosion

CASE HISTORY 1.1

Phenomenon: catastrophic general corrosion
Group/ Type: aluminum alloy, type AA5454 (UNS A95454)
Equipment: transport tankage
Environment: refined product
Conditions:
(a) major species: polar organic
(b) minor species: less than 0.2% water
(c) temperature: ambient to 167°F (75+°C) (during steam coil operation)
(d) pressure: atmospheric
(e) aeration: not given
Time to failure: less than 1 day
Illustration: current-time trace from scratch test
Comment: These aluminum storage tanks were completely penetrated after the product was heated with 302°F (150°C) steam. Estimated tank wall temperatures during steam heating were in excess of 167°F (75°C). Laboratory scratch tests of corrosion current density vs time showed that at temperatures of 122°F (50°C) and below, the passive film on the aluminum would reform. However, at 167°F (75°C) and above, the corrosion current density kept rising after the scratch, indicating that the passive film was unable to repair itself.
Remedy: Keep the tank wall temperatures to 122°F (50°C) maximum during steaming.

Reference: E.L. Liening, "Trouble-shooting Industrial Corrosion Problems with Electrochemical Testing Techniques," *Symposium on Non-Destructive Testing and Electrochemical Methods for Monitoring Corrosion in Industrial Plants* (Philadelphia, PA: ASTM 1984).

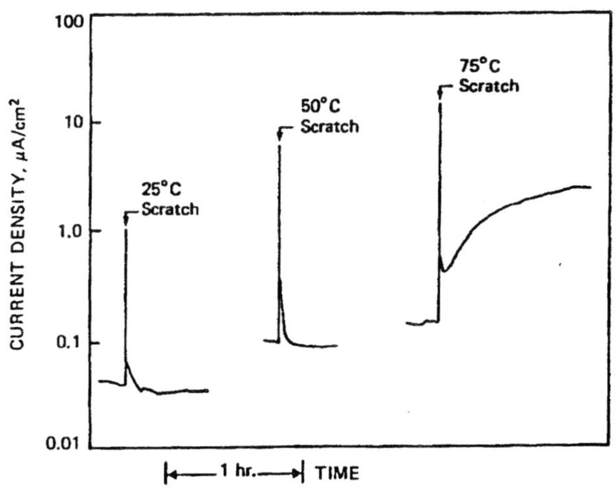

Figure 1.1 - Current-time trace from scratch test. (Reprinted with permission of ASTM.)

CASE HISTORY 1.2

Phenomenon: general corrosion
Group/ Type: ferrous, carbon steel
Equipment: air-cooled heat exchanger
Environment: plant air
Conditions:
(a) major species: air
(b) minor species: water, chlorides
(c) temperature: 136°F (58°C)
(d) pressure: atmospheric
(e) aeration: 20%

Time to failure: 12 years
Illustration: macro
Comments: Atmospheric corrosion of this finned air-cooled heat exchanger tube resulted in the loss of 0.030 in. of its original 0.075 inch wall over 12 years. Although this averages only 0.002 in./year, the loss was enough to produce a tensile failure from the longitudinal thermal stresses of startup, resulting in a major release of explosive process gas.
Remedy: Replace with thicker 14 gauge tubing, and coat the exposed portions with inorganic zinc paint.

Figure 1.2 - macro

CASE HISTORY 1.3

Phenomenon: external corrosion under insulation
Group/ Type: ferrous/ASTM A106 grade B (UNS K03006)
Equipment: piping, 3-in. schedule 40 piping
Environment: atmospheric, industrial, under insulation
Conditions:
(a) major species: water, air
(b) minor species: leachable chlorides
(c) temperature: 41°F (5°C) steady state, 212°F (100°C) during steam-out
(d) pressure: ambient
(e) aeration: saturated

Time to failure: 15 years
Illustration: macro
Comment: The temperature of this item was continuously below the atmospheric dew point, but not low enough to ice over the steel surfaces. Consequently, there was more or less continuous condensation of atmospheric moisture on the steel, which was not painted on installation. Corrosion under the foamglass insulation, which could normally be expected to progress at fairly low rates due to the low temperature, was aggravated by bimonthly steam-outs that caused any protective films formed to crack and spall. The result was unsuspected general corrosion at 14 mils/year, which eventually perforated this high-pressure gas line to start a major fire.
Remedy: Coat replacement piping with a high-build epoxy-phenolic maintenance paint, maintain the weather barrier, and check insulation periodically for moisture intrusion.
Contributor: W.G. Ashbaugh, Ashbaugh Corrosion Consulting Services, The Woodlands, TX.

Figure 1.3 - macro

CASE HISTORY 1.4

Phenomenon: general corrosion
Group/ Type: cast iron
Equipment: cannon
Environment: indoor atmosphere
Conditions:

(a) major species: air
(b) minor species: water vapor (60% relative humidity)
(c) temperature: indoor ambient 68 to 86°F (20 to 30°C)
(d) pressure: 1 atm
(e) aeration: saturated

Time to failure: 12 years
Illustration: photomacrograph, as-received and after 12 years on display
Comment: This cannon from a sunken Swedish warship sat on the sea bottom for 300 years and was reasonably well-preserved in the anaerobic mud on the ocean floor. Cast-in letters and a royal crest could still be read when the cannon was first recovered. However, after 12 years of display in an air-conditioned Stockholm museum, corrosion and destructive surface spalling obliterated all historically significant inscriptions. When recovered, all surfaces of the cannon were deeply embedded with hygroscopic chloride-bearing corrosion products. These products absorbed atmospheric moisture even in "dry" air-conditioned air, creating the conditions for self-perpetuating corrosion.
Remedy: Reduce all corrosion products back to elemental iron in a hydrogen furnace when such artifacts are first recovered, then seal with paraffin.
Reference: NBS publication, "Corrosion of Metal Artifacts," G. Ugiansky, Ed.

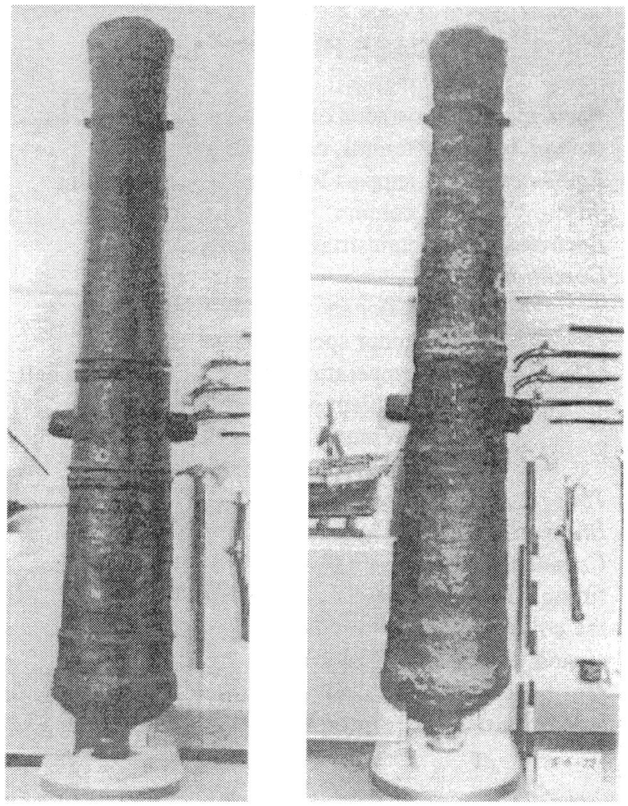

FIGURE 1.4 - (a) as received and (b) after 12 year's display.

CASE HISTORY 1.5

Phenomenon: general corrosion under insulation
Group/ Type: ferrous, carbon steel
Equipment: storage tank
Environment: industrial/marine atmosphere
Conditions:

(a) major species: air, water vapor
(b) minor species: NaCl, SO_2, H_2S
(c) temperature: ambient to 300°F (149°C)
(d) pressure: 1 atm
(e) aeration: restricted

Time to failure: not given
Illustration: Photograph
Comment: A spray-on urethane foam insulation was used to cut heat losses from an oil storage tank at a refinery. The oil was heated to 300°F (149°C) to keep it flowing. With time the insulation degraded, allowing water ingress to set up corrosion cells on the tank roof. In places, the tank roof had been penetrated by corrosion.
Remedy: Coat with epoxy-phenolic maintenance paint. Cover the insulation with SS weather barrier, inspect at two-year intervals for evidence of weather barrier or coating degradation.

CASE HISTORY 1.6

Phenomenon: general corrosion under insulation
Group/ Type: ferrous, carbon steel
Equipment: support beam on organic refining column
Environment: industrial atmosphere
Conditions:
(a) major species: air, water, vapor
(b) minor species:
(c) temperature: -45.5°C (-50°F) at shell, ambient outside insulation
(d) pressure: atmospheric
(e) aeration: saturated
Time to failure: 12 years
Illustration: photo
Comment: This T-beam was welded onto a butadiene refining column to support the overhead vapor line. Since the column operated continuously at -45.5°C, it was assumed there was no need for painting. However, this support beam protruded through the insulation out into the ambient air. Consequently, there was a temperature gradient across the beam from the cold column to the warm atmosphere, leading to continuous condensation of atmospheric moisture. This moisture produced corrosion at rates of about 35 mils/y, reducing this 0.5 in. thick beam to lacework in about 12 y.
Remedy: Coat the replacement with a high-build epoxy-phenolic maintenance paint.
Contributor: W.G. Ashbaugh

CASE HISTORY 1.7

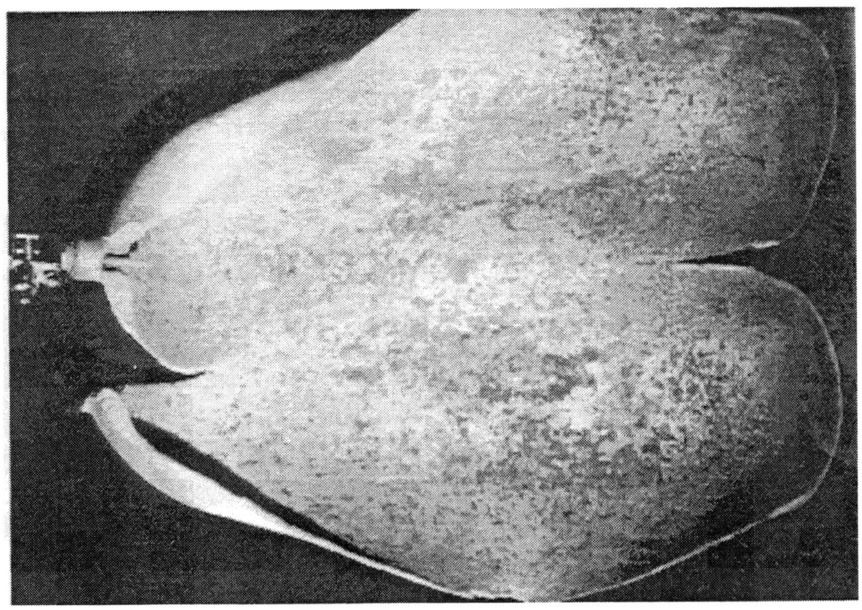

Phenomenon: general corrosion
Group/ Type: ferrous, galvanized carbon steel
Equipment: scuba tank
Environment: compressed air
Conditions:
(a) major species: air, water, vapor
(b) minor species: not given
(c) temperature: ambient to 49°C (120°F)
(d) pressure: 2,100 psi (143 atm)
(e) aeration: saturated
Time to failure: not given
Illustration: photo
Comment: This 3A steel scuba tank burst when pressurized to 2,100 psi, 85% of its rated working pressure. Subsequent analysis indicated water intrusion into the tank and water condensing from the compressed air had caused severe thinning of the tank wall. Upon rupture, the fracture propagated in a ductile manner into the upper and lower heads of the tank. The energy of the tank burst partially destroyed the dive shop and scattered debris as far as 400 ft.
Remedy: Use aluminum alloy scuba tanks.
Source: J.J. McAniff, National Underwater Accident Data Center, University of Rhode Island, Kingston, RI.

CASE HISTORY 1.8

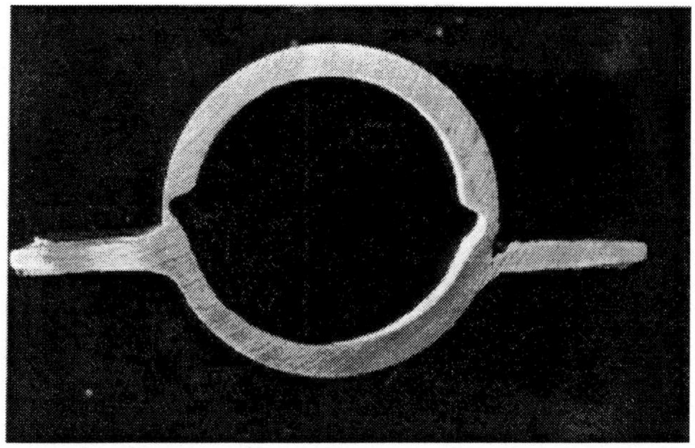

Phenomenon: general corrosion and "caustic gouging"
Group/ Type: ferrous, carbon steel
Equipment: roof tube in a boiler
Environment: boiler water and steam
Conditions:
(a) major species: water, steam
(b) minor species: traces of NaOH
(c) temperature: not given
(d) pressure: not given
(e) aeration: deaerated
Time to failure: 25 years
Illustration: macro of cross-section
Comment: Caustic and other dissolved solids may concentrate by evaporation at the water line in horizontal boiler tubes. The locally high caustic concentration leads to accelerated attack in parallel longitudinal trenches with nearly full thickness in nearby material exposed only to steam or the bulk solution. Such "caustic gouging" typically has smooth, rolling contours surrounded by an encrustation of boiler salts and, at times, black sparkling crystals of magnetite.
Remedy: not given
Reference: R.D. Port, Nalco Chemical Co.

CASE HISTORY 1.9

Phenomenon: general corrosion
Group/ Type: ferrous, carbon steel
Equipment: steam generator tube
Environment: boiler water
Conditions:
(a) major species: zeolite-softened water (hardness below 1.0 ppm, conductivity 1,500 to 2,000 micromhos)
(b) minor species: oxygen, chelant
(c) temperature: 500°F (260°C)
(d) pressure: 125 psi
(e) aeration: oxygen contaminated
Time to failure: total service life 25 years
Illustration: macro
Comment: Chelant corrosion in boilers occurs when the chelants concentrate by evaporation at water lines, or other locations where evaporating water leaves behind an accumulating residue of chelant. In the case illustrated above, excessive oxygen levels combined with chelant corrosion to produce sharply defined islands of intact metal surrounded by smooth plains of serious corrosion thinning. Such attack occurs most frequently in steam separation equipment.
Remedy: improved oxygen control
Reference: R.D. Port, Nalco Chemical Co., "Identification of Corrosion Damage in Boilers," CORROSION/84, paper no. 224 (Houston, TX: NACE, 1984).

CASE HISTORY 1.10

Phenomenon: general corrosion
Group/ Type: ferrous, carbon steel
Equipment: railroad tank car
Environment: concentrated sulfuric acid
Conditions:
(a) major species: 98% H_2SO_4
(b) minor species: not given
(c) temperature: ambient
(d) pressure: atmospheric
(e) aeration: not controlled
Time to failure: 3 years
Illustration: photograph
Comment: Ferrous sulfate films quickly cover the surface of steel exposed to stagnant, concentrated sulfuric acid and normally stifle further attack. However, in this tank car, hydrogen bubbles formed and floated up along the walls of the tank. The rising hydrogen bubbles eroded the ferrous sulfate film and produced renewed attack along a thin track which marked the bubbles' upward path. Ensuing bubbles would follow the same track and do the same sort of damage. After three years of service, thin grooves averaging 1/8 in. deep were cut into the tank walls from top to bottom.

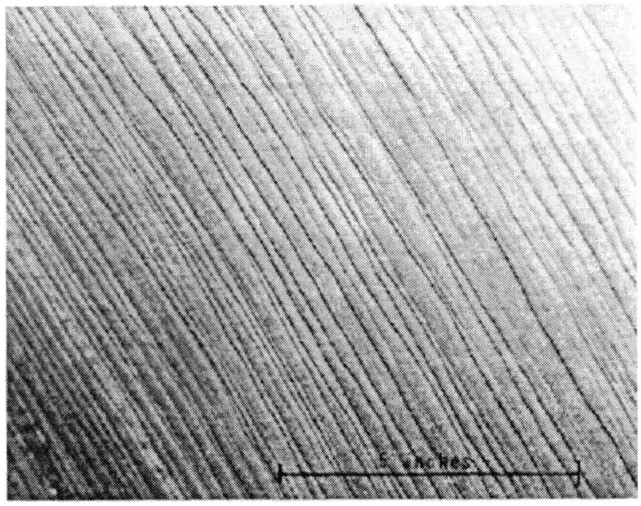

Remedy: Increase corrosion allowance.
Reference: S.L. Pohlman, R.V. Olson, "Corrosion and Materials Problems in the Copper Production Industry," CORROSION/84, paper no. 229 (Houston, TX: NACE, 1984).

CASE HISTORY 1.11

Phenomenon: general corrosion
Group/ Type: carbon steel, ASTM A214 gr. WCB
Equipment: valve for water injection line
Environment: Arabian aquifer water
Conditions:
a) major species: water, 19,204 mg/L Cl ion
b) minor species: 9,796 mg/L Na ion, 2,212 mg/L Ca ion, 1,061 mg/L SO_4 ion
c) temperature: 130ºF (54°C) max.
d) pressure: 100 psig
e) aeration: anaerobic
Time to failure: 8 years
Illustration: macro
Comment: For lack of an alternative, bare carbon steel valves were installed in a cement-lined steel pipeline handling corrosive aquifer water. The valve bodies typically perforated in six to eight years.

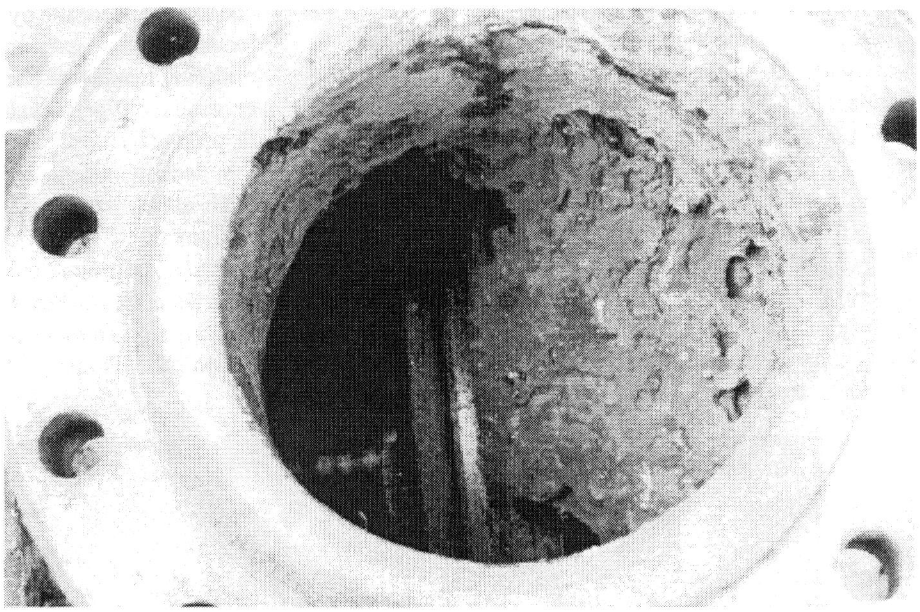

Figure 1.11 - Valve from Arabian aquifer service.

Remedy: Replace with CF3M stainless steel valves.

HISTORY 1.12

Phenomenon: general corrosion
Group/Type: ferrous, API 5L gr. B carbon steel
Equipment: sewer piping
Environment: oily water sewer
Conditions:

a) major species: water, hydrocarbons
b) minor species: H_2S, CO_2
c) temperature: ambient
d) pressure: 50 psig
e) aeration: unknown

Time to failure: 15 years
Illustration: photograph

Comment: General corrosion perforated this 12 inch, Sch. 40 line after 15 years' service carrying mixed waste streams from a natural gas processing plant. X-ray diffraction showed that the corrosion products in the piping consisted of 46% $FeCO_3$, 22% FeOOH and 19% iron sulfides of various compositions.
Remedy: Replace with cement-lined carbon steel.

CASE HISTORY 1.13

Phenomenon: general corrosion
Group/ Type: ferrous, carbon steel, 1 in. diameter ASTM A214 tubes
Equipment: air-cooled condenser
Environment: excess steam from a utility reboiler
Conditions:

a) major species: steam and steam condensate, nominal pH 8.7 to 9.2
b) minor species: traces of H_2S and CO_2
c) temperature: 215ºF (101°C) inlet, 180ºF (82°C) outlet
d) pressure: 70 psig inlet, 60 psig outlet
e) aeration: unknown

Time to failure: 15 years

Illustration: photograph
Comment: The condenser cools excess steam off a utility reboiler, which boils out rich di-glycol amine containing H_2S and CO_2. The tube failed by general corrosion when its thickness was reduced below that needed to support the mechanical load. X-ray diffraction analysis indicated that the corrosion products were largely oxides of iron, but they also contained 3% pyrrhotite (Fe1-xS), 1% siderite ($FeCO_3$) and 1% $CaCO_3$, indicating contamination of the steam by acid process gas from the reboiler.
Remedy: Eliminate tube and tubesheet leaks in the reboiler.

CASE HISTORY 1.14

Phenomenon: general corrosion
Group/ Type: ferrous, carbon steel, API 5CT grade P110
Equipment: oil well casing
Environment: external, desert sand
Conditions:
a) major species: sand, water
b) minor species: NaCl, carbonates, sulfates
c) temperature: ambient
d) pressure: atmospheric
e) aeration: saturated

Time to failure: 10 years
Illustration: photograph
Comment: To hold back drifting desert sand during completion, several dozen Arabian oil wells were drilled and completed with carbon steel boxes built around the upper end of the casing and wellhead assemblies, as shown

on the photograph. Once brought on production, these boxes shielded the upper end of the casing and the wellhead from the impressed current of the cathodic protection system. After some years on production, the casing perforated from the outside due to soil side corrosion.
Remedy: Attach two small sacrificial anodes to the casing and wellhead inside the box.
Contributor: D.R.Catte, Saudi Aramco

CASE HISTORY 1.15

Phenomenon: general corrosion
Group/ Type: ferrous/carbon steel
Equipment: storage tank
Environment: solvent/water mixture
Conditions:
(a) major species: 65% H_2O, 25% acetone, 3 to 5% isopropanol
(b) minor species: 1.5% vinyl acetate
(c) temperature: ambient
(d) pressure: atmospheric
(e) aeration: saturated

Time to failure: 8 years
Illustration: none
Comments: Due to traces of acetic acid liberated by the vinyl acetate, the pH of this water-solvent mixture averaged ~5. The bottom and lower shell of the tank corroded quite slowly over the tank's life at rates averaging 3 mpy (0.076 mm/year). However, the roof corroded at 47 mpy (1.19 mm/year), apparently resulting from a higher concentration of acetic acid in droplets condensing on the roof.
Remedy: Replace the damaged portions with AISI 304 SS (UNS S30400).

CASE HISTORY 1.16

Phenomenon: general corrosion
Group/ Type: ferrous carbon steel
Equipment: reboiler tubes
Environment: steam condensate
Conditions:
(a) major species: water
(b) minor species: 20 ppm CO_2
(c) temperature: 212 to 302°F (100 to 150°C)
(d) pressure: 700 psi
(e) aeration: not determined

Time to failure: 2 years
Illustration: none
Comments: Steel tubes in a reboiler failed from steam-side condensate after two years. At the steam inlet end, corrosion was a relatively mild 5 mpy (0.13 mm/y). However, at the condensate outlet end of the exchanger, CO_2 in the condensing steam lowered the pH and raised the corrosion rate to 30 mpy (0.76 mm/y).
Remedy: Eliminate CO_2 with stream traps or retube with AISI 316 SS (UNS S31600).

CASE HISTORY 1.17

Phenomenon: general corrosion
Group/ Type: ferrous, carbon steel
Equipment: piping
Environment: liquid chlorine
Conditions:

(a) major species: chlorine
(b) minor species: 5 to 30 ppm H_2O, 3.5% HCl
(c) temperature: ambient
(d) pressure: not specified
(e) aeration: not determined

Time to failure: ~6 months
Illustration: schematic
Comment: Two tanks containing chlorine from different sources were tied together to a common outlet line. One tank contained HCl-free chlorine with ~30 ppm H_2O. The other tank contained anhydrous (below 5 ppm water) Cl_2 with 3.5 mol% HCl. The outlet line failed by general thinning within a few months, causing a major chlorine release.
Remedy: Run a separate outlet line from each tank.
Reference: T.F. Degnan, "Unusual Failures in Chlorine Environments," CORROSION/83, paper no. 22 (Houston, TX: NACE, 1983)

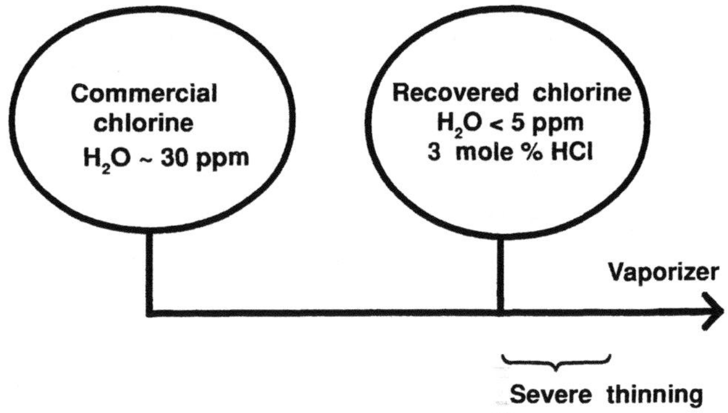

CASE HISTORY 1.18

Phenomenon: general corrosion
Group/ Type: ferrous, carbon steel and alloy 20Cb3
Equipment: condenser tubes
Environment: shell-side $CaCl_2$ brine, tube-side Cl_2 + HCl
Conditions:

(a) major species: chlorine
(b) minor species: HCl, 1 to 2 ppm H_2O
(c) temperature: 32°F (0°C) brine temperature
(d) pressure: not specified
(e) aeration: not determined

Time to failure: less than three months
Illustration: schematic, solubility graph
Comment: HCl in chlorine greatly lowers its solubility for water. When the solubility limit is exceeded, a separate water phase is formed, which is rich in HCl and chlorine. This condenser lost alloy 20Cb3 tubes quickly because of this highly corrosive water phase and was retubed in heavy-walled carbon steel. The carbon steel tubes began leaking in a few weeks. When the unit was shut down, chlorine and HCl leaked into the shell-side of the condenser and generated enough hydrogen on the wet steel to cause a major plant fire.
Remedy: Raise the brine temperature enough so that the traces of water (1 to 2 ppm) remain in solution in the chlorine.
Reference: T.F. Degnan, *Materials Performance,* 22, 11 (1983): p. 45-47.

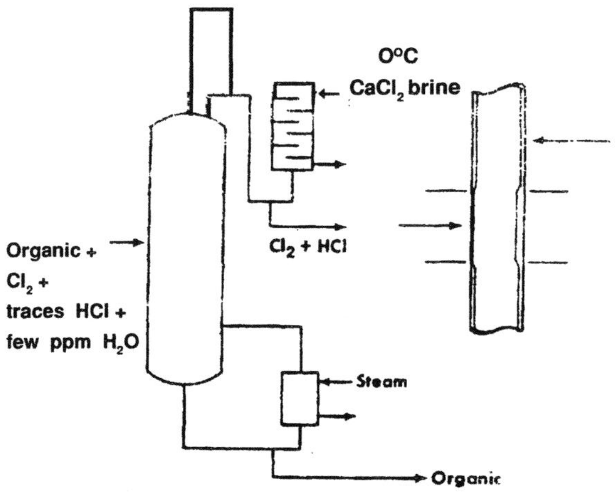

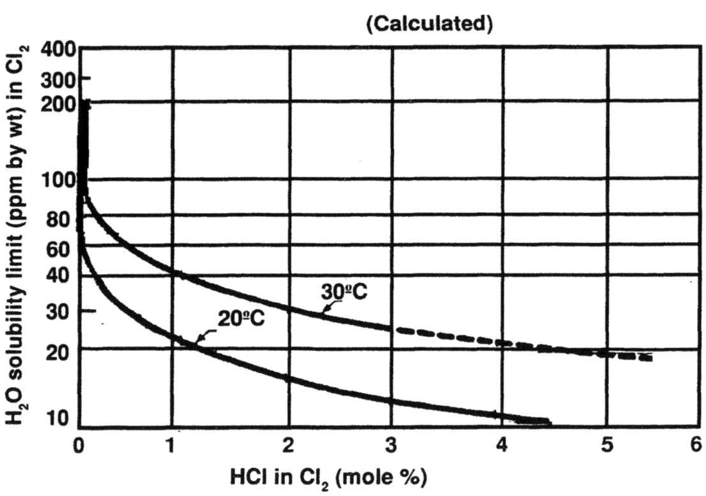

CASE HISTORY 1.19

Phenomenon: general corrosion
Group/ Type: ferrous, carbon steel
Equipment: condenser tubes and shell
Environment: crude oil distillate
Conditions:
(a) major species: low-boiling hydrocarbons
(b) minor species: HCl (2,000 ppm Cl-, H_2S)
(c) temperature: 248°F (120°C) inlet, 185°F (85°C) outlet
(d) pressure: 1 atm
(e) aeration: anaerobic
Time to failure: 10 months
Illustration: schematic
Comment: Overhead condensers for atmospheric distillation of crude oil form HCl from chloride salts (if desalting is not complete) and H_2S from sulfur compounds. This strongly acid shell side quickly corrodes carbon steel.

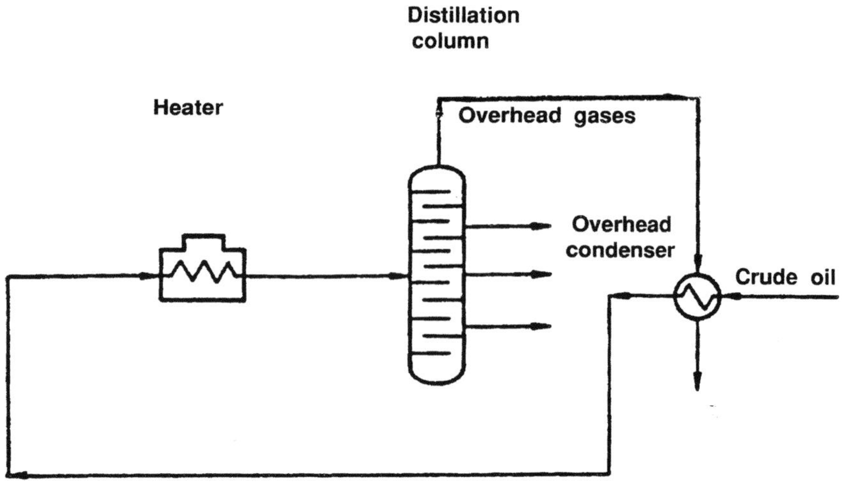

Remedy: UNS N08904 SS (20 Cr, 25 Ni, 4.5 Mo) replacement tubes in service seven years with no problems.
Reference: C. Martenson, G. Supko, "Experience with High Alloy Stainless Steels in the HPI," *Hydrocarbon Processing,* 1 (1983): p. 69-74.

CASE HISTORY 1.20

Phenomenon: general corrosion
Group /Type: carbon steel
Equipment: heat exchanger tubes
Environment: ethylene dichloride reactor
Conditions:
(a) major species: liquid ethylene dichloride
(b) minor species: 1,000 ppm HCl, 25 ppm $FeCl_3$, several hundred ppm Cl_2; low water (ppm)
(c) temperature: 151°F (66°C)
(d) pressure: not given
(e) aeration: "a few hundred" ppm O_2
Time to failure: not given
Illustration: Schematic
Comment: Nonpolar organic solvents are, when dry, normally considered incapable of supporting the electrochemical reactions that produce corrosion in aqueous streams. However, in these coolers, metal loss ranged from 23 to 45 mpy (0.58 to 1.1 mm/year) at tube inlets. The penetration rate was progressively lower further into the tubes. Investigation showed the operation of a low temperature direct reaction between chlorine and steel. The reaction rate could be stifled by the presence of $FeCl_3$ corrosion products on the steel surfaces. When these were swept away by turbulence near tube inlets, wall-thinning by this unusual corrosion mechanism proceeded to the point of perforation.

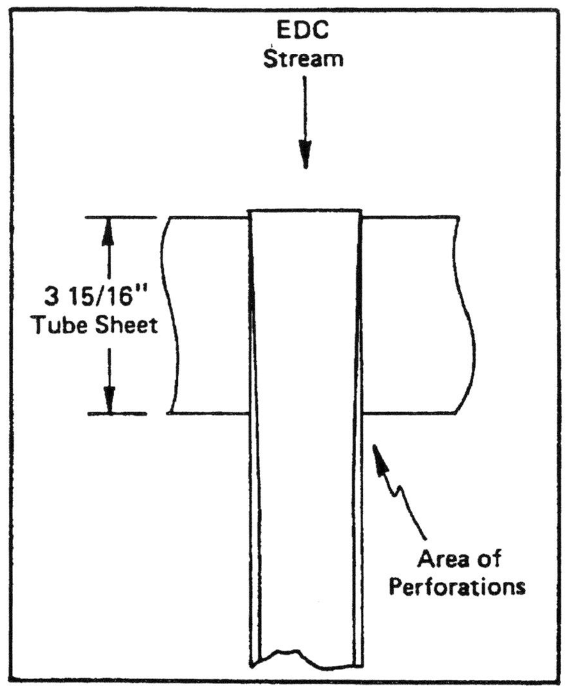

Remedy: Use protective inserts over the tube inlet ends.
Reference: W.C. Fort III, B.T. Ellison, "Accelerated Entrance Corrosion of Steel Tubes by Chlorine in a Nonaqueous Liquid at Low Temperature," CORROSION/83, paper no. 14 (Houston, TX: NACE, 1983)

CASE HISTORY 1.21

Phenomenon: general corrosion
Group/ Type: carbon steel, ASTM A106 gr. B, Sch. 80
Equipment: piping
Environment: refinery slop wax
Conditions:
a) major species: heavy hydrocarbons
b) minor species: 2 wt% sulfur compounds
c) temperature: 550ºF (288°C)
d) pressure: 150 psig
e) aeration: anaerobic
Time to failure: 6 years
Illustration: macro

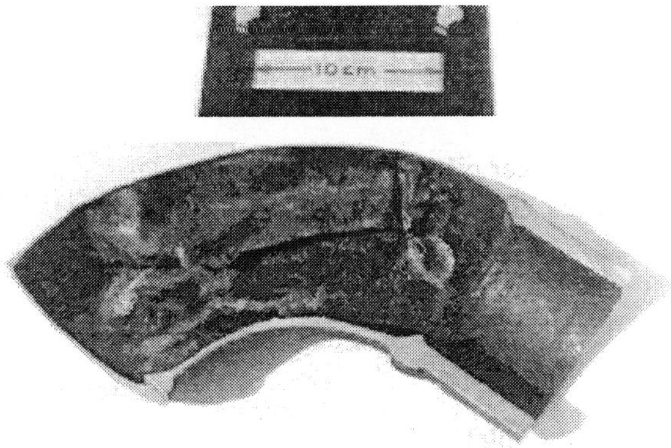

Figure 1.21- Cross section of corroded carbon steel elbow. Note full thickness on 5 Cr steel pipe above the elbow.

Comment: Slop wax is a heavy sulfur-bearing hydrocarbon, the bottoms off a crude oil refinery's vacuum column. It is corrosive to carbon steel at the service temperature of 550ºF (288°C), although 5Cr steels show almost nil corrosion. By mistake this carbon steel pipe elbow was welded into a 5Cr steel piping run. After six years service the Schedule 80 4-in. elbow was thinned to perforation by general corrosion, causing a major fire. Note that the 5Cr pipe to which the elbow was welded still showed original thickness.
Remedy: Determine the alloy on all pipes and fittings in the slop wax line using chemical spot tests and portable x-ray fluorescence analyzers. Replace all carbon steel pipes and fittings with 5Cr steel.

CASE HISTORY 1.22

Phenomenon: general corrosion
Group/Type: carbon steel, API 5L gr. B
Equipment: molecular seal on flare tip
Environment: acid gas from gas processing plant
Conditions:
a) major species: methane
b) minor species: 5-8% H_2S, 2-6% CO_2, water vapor
c) temperature: 450ºF (232°C) nominal
d) pressure: 2-4 psig
e) aeration: intermittent
Time to failure: six years
Illustration: macro

Comment: This molecular seal is a steel cap which fits over the end of the flare line to prevent air getting into the flare line and forming an explosive mixture. The flare operates intermittently; when it is not burning, a purge of 60 psig steam is maintained through the line. The 13 mm steel plate of the seal perforated due to external general corrosion after six years' service. The alternate exposure to the H_2S-bearing flared gas during flare operation and oxygenated water from moist air and condensed steam during periods of low or no flow of the flared gas prevented formation of a protective sulfide film on the steel and caused the relatively high corrosion rate of 90 mils per year.
Remedy: Coat the repaired seal with an inorganic zinc coating, reduce condensation by increasing the purge gas flow rate and steam temperature.

CASE HISTORY 1.23

Group/ Type: ferrous, ASTM A214 carbon steel tubing, 12 gauge
Phenomenon: general corrosion
Equipment: air-cooled heat exchanger tubes
Environment: crude oil distillation overhead stream
Conditions:

a) major species: hydrocarbons (gasoline, naptha, bunker oil)
b) minor species: water, 60 ppm Cl-, H_2S
c) temperature: 128ºF (53°C)
d) pressure:
e) aeration: anaerobic

Time to failure: 4 years
Illustration: photograph
Comments: The photograph shows the inlet end of an air-cooled HX tube, split longitudinally, with one half sand-blasted to remove the corrosion products. Classic refinery dew point corrosion was observed, in which concentrated HCl in the first fluid to condense causes severe general attack on the first 6 to 12 inches of the tube's length, reducing the thickness to a feather edge at the inlet end. This unit had operated for some time with pH swings of 3 to 8 in the overhead accumulator water, with chloride content in the accumulator water averaging 60 ppm.

Figure 1.23 - Photomacrograph of cross-section

Remedy: Improve crude oil desalting to reduce the chloride content in the overhead accumulator water to 10 to 30 ppm, and inject NaOH and neutralizing amines to control the accumulator water pH between 5.5 and 6.5.

CASE HISTORY 1.24

Phenomenon: General corrosion and hydrogen induced cracking
Group/ Type: carbon steel, API 5L gr X-52
Equipment: acid gas flare header
Environment: waste natural gas
Conditions:

a) major species: methane
b) minor species: 5% H_2S, 3% CO_2, water-saturated
c) temperature: ambient (40-120ºF)
d) pressure: 0-5 psig operation, 40 psig design
e) aeration: intermittent

Time to failure: 12 years
Illustration: macro of cross-section
Comment: This 36-inch header piped waste gas from a natural gas processing plant to a flare. The flare line operated at low pressure with occasional periods of low or no flow, which made it prone to occasional ingress of atmospheric air and moisture. After 12 years severe general corrosion was noted on ultrasonic testing. HIC was observed in many areas which showed only moderate general corrosion.

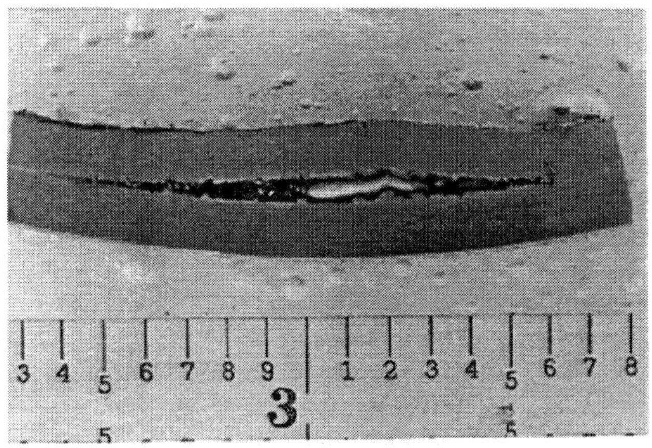

Remedy: Purge the line with dry fuel gas during low flow periods, replace the damaged sections with HIC-resistant steel of low sulfur content with calcium additions for inclusion shape control.

Figure 1.25 - Sulfur pit sump piping.

CASE HISTORY 1.25

Phenomenon: general corrosion
Group/ Type: ferrous, carbon steel
Equipment: steam coil piping and hardware
Environment: molten sulfur pit
Conditions:

a) major species: molten sulfur
b) minor species: water with pH = 1.48, 0.4% H_2SO_4, 2,000 mg/L Fe
c) temperature: 280ºF (138°C) steam temperature
d) pressure: liquid head only
e) aeration: anaerobic

Time to failure: 1 year
Illustration: photograph
Comment: Sulfur in a storage pit was kept liquid by heat from a steam coil in the bottom of the pit. This steam coil and its supports and clamps were in good condition except near the sump, the lowest portion of the pit, where wash water and ground water could collect. On analysis, the liquid in the sump showed a pH of 1.48, severely corrosive to carbon steel.
Remedy: Replace the steel sump piping with alloy 20 or 904L.

CASE HISTORY 1.26

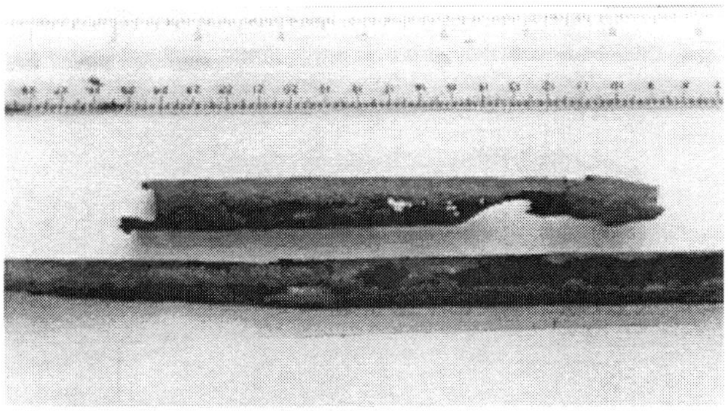

Figure 1.26

Phenomenon: general corrosion
Group/Type: ferrous, carbon steel
Equipment: crude oil preheater
Environment: crude oil
Conditions:

a) major species: 75% API gravity 33.8 crude oil, 24% atmospheric distillation column bottoms
b) minor species: 0.14% water and sediment, 4.3 lbs/1,000 barrels salt, 1.77 wt% total sulfur
c) temperature: shell side inlet 100ºF (37.8°C), tube side inlet 610ºF (321°C)
d) pressure: 400 psig shell side
e) aeration: anaerobic

Time to failure: 1 year
Illustration: photograph
Comment: These ¾-in. diameter, 0.083 in. wall thickness carbon steel tubes were perforated by shell side corrosion between the tubesheet and the first baffle after only one year's service. Corrosion products were largely iron sulfides with some $CaCO_3$.
Remedy: Replace the steel tubes with nickel-based high-chromium, high-molybdenum alloy.

CASE HISTORY 1.27

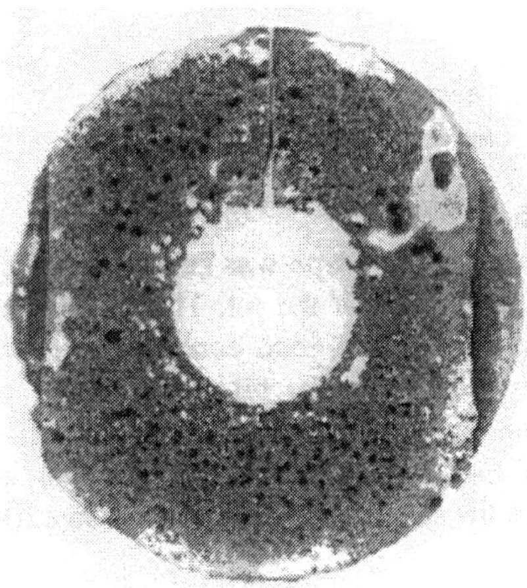

Figure 1.27(a)

Figure 1.27(b)

Phenomenon: general corrosion with end grain attack
Group/ Type: stainless steel, AISI 316 (UNS S31600)
Equipment: plunger valve seat
Environment: polyvinyl chloride
Conditions:

(a) major species: PVC, steam
(b) minor species: hydrochloric acid, hydrogen peroxide
(c) temperature: 59°F (15°C) steady state, 176 to 230°F (80 to 110°C) on steam-out
(d) pressure: 25 psig steady state, ambient on steam-out
(e) aeration: intermittent

Illustrations: macros
Comments: Periodically, this item was steam-cleaned to remove caked resin adhering to the walls. The steam-out hydrolyzed some of the PVC to dilute hydrochloric acid, thus causing intermittent but severe corrosion of the SS vessel internals. This AISI 316 SS (UNS S31600) valve seat was machined out of bar stock, thus leaving the short-transverse direction, the end-grain microstructure, exposed. After only 2,500 total hours of operation, and two steam cleanings, general corrosion was observed with end grain attack proceeding at rates of ~7 in. (175 mm)/y.
Remedy: Change to UNS N04400 alloy.

CASE HISTORY 1.28

Phenomenon: general corrosion
Group/ Type: stainless steel, AISI 304(UNS S30400)
Equipment: piping
Environment: vent scrubber effluent
Conditions:
- (a) major species: methyl ethyl ketone (MEK), water
- (b) minor species: vinyl chloride, vinyl acetate, HCl, pH 1
- (c) temperature: 140°F (60°C)
- (d) pressure: not given
- (e) aeration: not given

Time to failure: not given
Illustration: none
Comment: MEK solutions are normally non-corrosive to SS. However, when heated above ~140°F (60°C), vinyl chloride can hydrolyze to produce HCL. The acid is only slightly soluble in MEK and thus concentrates in the water phase, giving HCL strengths much higher than would be expected based solely on the bulk weight percentage of vinyl chloride. The result is extremely high corrosion rates along the bottom of the piping, where the water phase collects.
Remedy: Dehydrate, adjust pH, or use NiCrMo alloys.

CASE HISTORY 1.29

Phenomenon: general corrosion
Group/ Type: ferrous, 26Cr-1Mo SS
Equipment: heat exchanger tubing
Environment: chlorinated organic
Conditions:
- (a) major species: water, ammonia, amine hypochlorite
- (b) minor species: not determined
- (c) temperature: 284°F (140°C)
- (d) pressure: 100 psi
- (e) aeration: not given

Time to failure: 10 days
Illustration: macro

Comment: This reactor feed preheater was tubed in 26-1 SS based on field corrosion studies which showed low rates in this process. During unit startup this bundle was on standby with steam circulating on the shell side. The process side was inadvertently valved in, leaving stagnant reactor feedstock in the tubes. The stagnant feedstock broke down to make concentrated HCl in the tubes. Perforation of the 0.065-in. wall thickness tubes in five days was the result.
Remedy: Revise start-up procedure to circulate steam condensate through the tubes during standby periods.

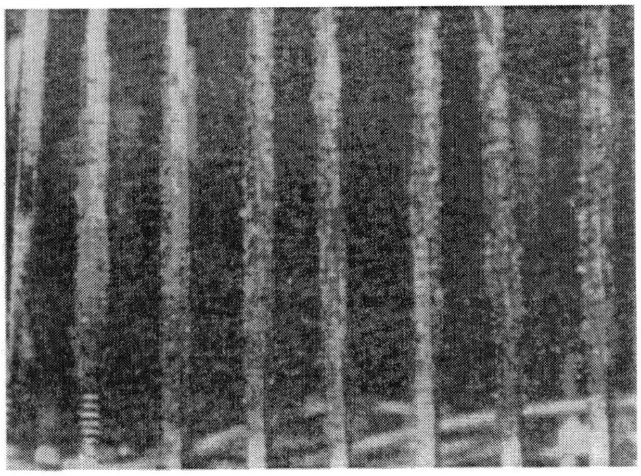

Figure 1.30

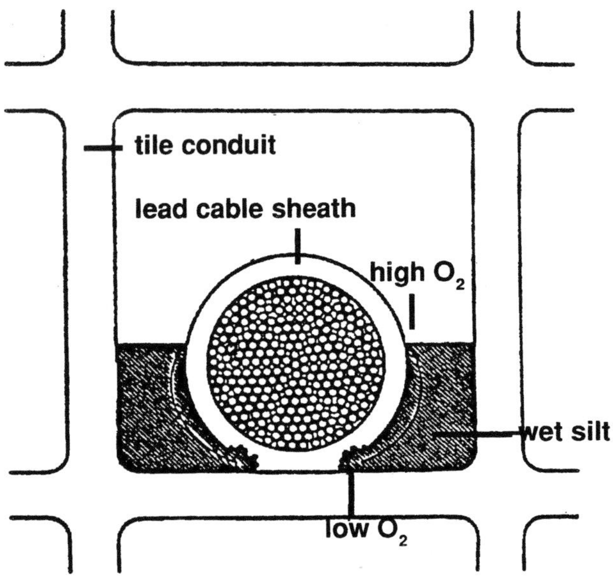

Figure 1.31

CASE HISTORY 1.30

Phenomenon: general corrosion
Group/ Type: copper and lead-tin alloy foils
Equipment: electronic printed circuit boards
Environment: ambient atmosphere, newspaper composing room
Conditions:
(a) major species: air, water vapor
(b) minor species: vapors of KOH, HAc, sodium thiosulfate and methylformate in tap water
(c) temperature: ambient
(d) pressure: 1 atm
(e) aeration: saturated

Time to failure: 6 months
Illustration: macro
Comment: A typesetter in an enclosed air-conditioned newspaper office failed due to corrosion of the printed circuit board connections. The typesetter's cooling fan had pulled fumes from a nearby photographic developing and fixing station into the typesetter casing. The fumes attacked the connections to the typesetter control board, rendering it inoperable.
Remedy: Typesetter problems ceased once the fumes from the photographic developing operation were captured and vented to the outside.
Reference: B.J. Boan, "Corrosion of Printed Circuit Boards by Photographic Chemical Vapors," CORROSION/84, paper no. 40 (Houston, TX: NACE: 1984)

CASE HISTORY 1.31

Phenomenon: general corrosion (differential aeration)
Group/ Type: lead
Equipment: cable sheathing
Environment: wet silt
Conditions:
(a) major species: water, silt
(b) minor species:
(c) temperature: ambient
(d) pressure: atmospheric
(e) aeration: 0 to saturated (see figure)

Time to failure: not given
Illustration: drawing
Comment: This cable was pressed tightly against the inner surface of a tile duct. The area against the tile did not corrode. However, adjacent areas, where oxygen was depleted, corroded in preference to the lead closer to the surface, which had more air available for film repair.
Remedy: Exclude water and silt from the cable duct.
Reference: *Corrosion*, 12, 5 (1956): p. 257t-259.

CASE HISTORY 1.32

Phenomenon: general corrosion
Group/ Type: copper alloys: silicon bronze (UNS C65500)
Equipment: corrosion coupon
Environment: phosphoric acid
Conditions:
(a) major species: 56% phosphoric acid
(b) minor species: not given
(c) temperature: 550°F (288°C)
(d) pressure: 900 psi
(e) aeration: anaerobic
Illustrations: macro, micro
Comments: The illustrations show a weight-loss coupon used to monitor corrosion in a process plant. As droplets of acid drip down along the surface of the coupon, general corrosion removes copper from the center of the coupon at local anodes. At the same time, copper is plating out along the bottom edge of the coupon, apparently at the site of a local cathode. The result is a curious redistribution of material with very little net weight loss.
Remedy: Increase corrosion allowance.

Figure 1.32(a)

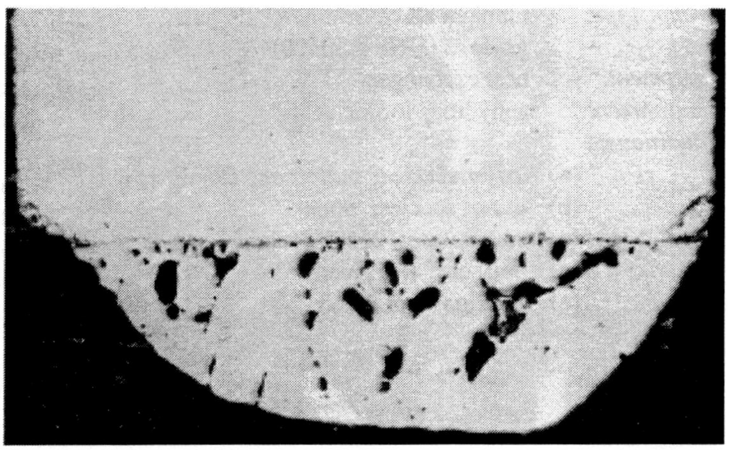

Figure 1.32(b) - Cross section 50X.

CASE HISTORY 1.33

Phenomenon: general corrosion
Group/ Type: copper alloys/admiralty brass (UNS C44300)
Equipment: condenser tubes
Environment: sour gas processing
Conditions:
(a) major species: H_2S, H_20,
(b) minor species: diethyleneamine
(c) temperature: 248°F (120°C)
(d) pressure: not given
(e) aeration: anaerobic
Time to failure: 1 year
Illustration: none
Comments: Because of time constraints during a maintenance outage, 0.065-in. wall admiralty brass heat exchanger tubes were used to replace carbon steel in a sour refinery gas condenser. The inhibited H_2S-bearing condensate, which gave tolerably low corrosion rates on steel, perforated the brass tubes within a year. Amine inhibitors will not protect copper alloys in sour gas environments and may, in fact, serve as pitting or SCC agents due to the liberation of ammonia.
Remedy: Replace with ferritic SS.

CASE HISTORY 1.34

Phenomenon: general corrosion
Group/ Type: nickel-based alloys; alloy 400 (UNS N04400)
Equipment: product transfer hose
Environment: 20% ferric sulfate
Conditions:
(a) major species: water, ferric sulfate
(b) minor species: chloride (20 ppm), oxygen
(c) temperature: 86°F (30°C)
(d) pressure: 20 psi
(e) aeration: saturated
Time to failure: 3 to 6 hours
Illustration: none
Comment: A ferric sulfate solution was being transferred from a storage tank to a tank truck. In the course of an 8-hour shift, three corrugated-metal flexible transfer hoses were connected to the tank in succession, then replaced as each began to leak. Spot tests indicated that the hoses were alloy 400 rather than AISI 304 SS (UNS S30400) as assumed. Nickel-copper alloy 400 has good resistance to many deaerated reducing acids. However, a 20% ferric sulfate solution behaves similarly to dilute sulfuric acid with heavy concentrations of the oxidizing ferric ion. The resistance of alloy 400 to oxidizing acids is generally poor even at room temperature.
Remedy: Spot test transfer hoses for alloy before use; use AISI 304 SS hoses for ferric sulfate transfer.

CASE HISTORY 1.35

Phenomenon: general corrosion
Group/ Type: titanium alloy, commercially pure, grade 2 (UNS R50400)
Equipment: heat exchanger
Environment: anhydrous formic acid
Conditions:
(a) major species: anhydrous formic acid
(b) minor species: none
(c) temperature: 266°F (130°C)
(d) pressure: not given
(e) aeration: not given
Time to failure: 2 days (six months total service)
Illustration: none
Comment: For six months, titanium tubes successfully resisted the acetic acid-25% formic acid process for which this exchanger was selected. However, after a unit turnaround the tubes were perforated in two days when refined, anhydrous formic acid was circulated during startup. Titanium has good resistance to oxidizing acids, but poor resistance to reducing acids, such as pure formic.
Remedy: Circulate acetic/formic mixture during startup..

CHAPTER 2

Localized Corrosion

CASE HISTORY 2.1

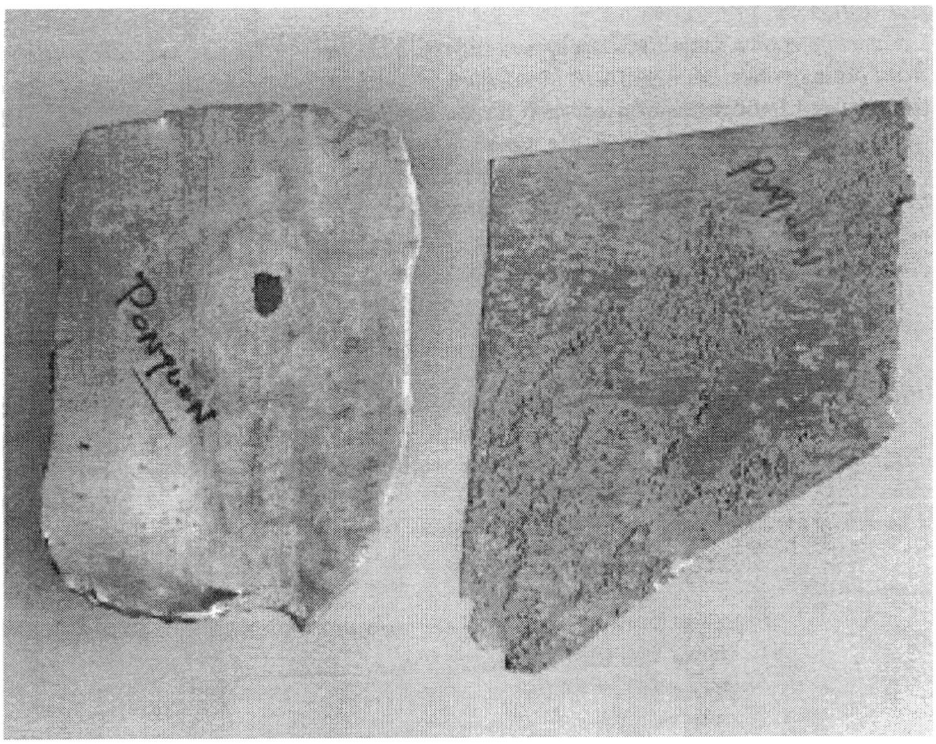

Phenomenon: pitting corrosion
Group/ Type: aluminum alloys 2024 (UNS A92024) and 5005 (UNS A95005)
Equipment: storage tank floating roof
Environment: anhydrous methanol
Conditions:
(a) major species: methanol
(b) minor species: 0.04% H_20, 0.002% HAc, 20 ppm NaOH
(c) temperature: ambient
(d) pressure: atmospheric
(e) aeration: not deter mined
Time to failure: total life eight years
Illustration: macro
Comment: Anhydrous (less than 0.1% H_20) alcohols with less than five carbon atoms occasionally produce catastrophic pitting corrosion of aluminum alloys. The cause of this attack is not well understood, but the attack may be extremely rapid with voluminous hydrogen evolution and destructive metal loss after an unpredictable initiation time ranging from a few days to several thousand hours. The aluminum pontoons on a floating roof tank sank after several days storage of a shipment of methanol that was particularly low in water. The caustic content in this particular shipment, which was about 10 times normal, apparently served to help depassivate the aluminum surfaces and initiate the attack.
Remedy: Do not store anhydrous ethanol, methanol or isopropanol in aluminum equipment for long periods. (Relatively brief containment in aluminum tank trucks has been commercially successful.)

CASE HISTORY 2.2

Phenomenon: pitting
Group/ Type: ferrous/AISI 1008 (UNS G10080) carbon steel
Equipment: domestic water piping; 1 in. diam, 0.086 in. wall
Environment: city well water
Conditions:
(a) major species: water
(b) minor species: 6 ppm CO_2, 44 ppm Cl-, 292 ppm bicarbonate alkalinity
(c) temperature: 95 to 104°F (35 to 40°C)
(d) pressure: 40 psig
(e) aeration: 3.75 ppm dissolved O_2
Time to failure: 22 to 36 months
Illustration: macro
Comments: A building's hot water system suffered 17 leaks from pitting within 36 months of installation. Examination showed light deposition of scale on the inside surfaces of the pipes. Large tunnel-like tubercles (nodules of iron oxide and iron carbonate) had formed over local sites of attack. The galvanized coating was largely intact except under the tubercles. Approximately half the leaks occurred at rough ERW seams, with heavy attack in the weld metal. The rest of the pits were randomly distributed. The pits were attributed to dissolved oxygen in the water. Potential reversal between the zinc coating and the steel was postulated because of the oxygen and bicarbonate levels present. This possibility was not, however, established conclusively.
Remedy: Raise the water pH to promote a thicker, more adherent scale and use a food-grade phosphate inhibitor. Use copper or FRP pipe on new construction.

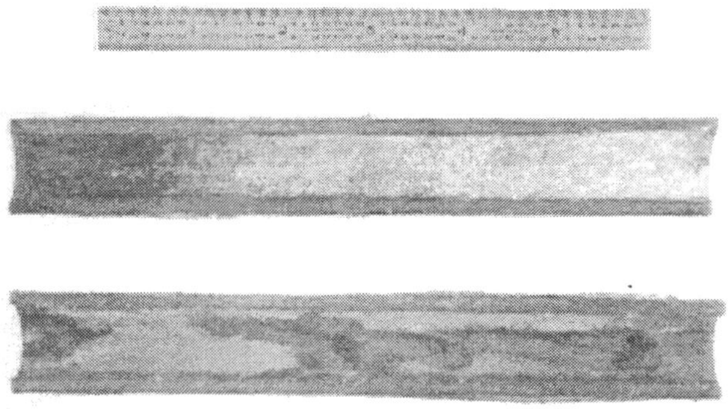

CASE HISTORY 2.3

Phenomenon: pitting
Group/ Type: ferrous, carbon steel piping
Equipment: 14 in. diameter, 0.375 in. wall thickness piping
Environment: softened chilled water to an air conditioning plant
Conditions:
a) major species: water
b) minor species: 100 ppm chlorides, 90 ppm sulfates
c) temperature: 50ºF (10°C)
d) pressure 50 psig
e) aeration: 2 ppm O_2
Time to failure: 10 years
Illustration: photograph
Comment: Corrosion products in the base of the pit consisted largely of Fe_3O_4 and FeOOH with 2% Fe_9S_8. It was concluded that the primary cause of the pitting attack was oxygen with some activity by sulfate-reducing bacteria to produce the iron sulfides. Although SRBs thrive only in anaerobic environments, oddly enough, the most severe cases of bacterial corrosion occur in partly aerated systems, where aerobic bacteria form slimes beneath which the anaerobic SRBs can take shelter.
Remedy: Treat the water with chemical oxygen scavengers and biocides.

CASE HISTORY 2.4

Phenomenon: pitting
Group/ Type: ferrous/API 5LX-52 carbon steel line pipe
Equipment: gas transmission piping; 0.25 in. wall
Environment: ambient air, soil
Conditions:

(a) major species: air, silica sand
(b) minor species: water vapor, chlorides, sulfides,
(c) temperature: ambient (10 to 35°C)
(d) pressure: atmospheric
(e) aeration: variable at crevices

Time to failure: 16 to 22 weeks
Illustration: macros
Comments: Steel pipe with surfaces contaminated with sulfate compounds from an unknown source was stored outdoors on sand berms in a yard ~20 miles from the Gulf of Mexico. After 16 to 20 weeks, up to 8% of the pipe joints had pits exceeding 5% of the wall thickness. The deepest pit observed was 0.086 in. (2.18 mm) after 20 weeks of storage. About 90% of the pits occurred in areas where damp sand touched the bottom of the pipe. At such points, reddish-brown stains signaled the presence of pits beneath the sand (Figure). Although the sand was free of sulfates, sulfides were detected in the pits both with the sodium azide spot test and with EDAX. Cultures of sand samples for sulfate-reducing bacteria (SRB) were positive. Some shallow, insignificant pits occurred on the edges of bands of black oxide at the lines of contact between neighboring joints of pipe in the stack.
Remedy: Store the pipe on raised racks of concrete beams or railroad ties to prevent contact with the damp sand.

Figure 2.4(a)

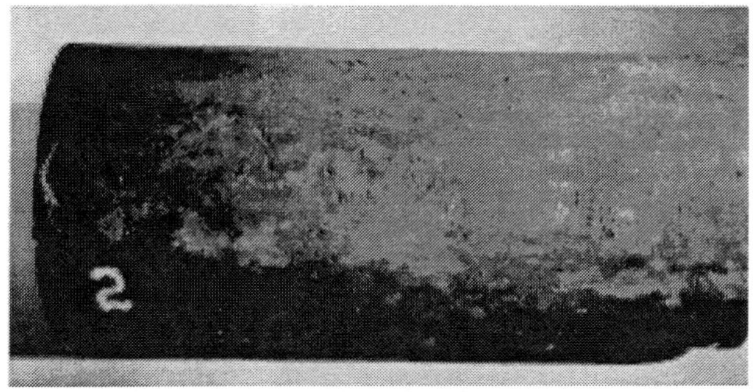

Figure 2.4(b)

CASE HISTORY 2.5

Phenomenon: pitting
Group/ Type: ferrous/API 5LX-52 carbon steel line pipe
Equipment: gas transmission line
Environment: sweet wellhead gas
Conditions:

(a) major species: methane
(b) minor species: 1 to 3% CO_2, saturated with H_2O vapor
(c) temperature: 105°F (41°C)
(d) pressure: 800 psig
(e) aeration: anaerobic

Time to failure: 20 years
Illustration: macro and cross-section
Comment: This carbon steel gas transmission pipeline was inhibited with an oil-soluble, water-dispersible inhibitor. After 20 years service, the line pipe showed negligible corrosion, but many of the welds were pitted to perforation. All the attack occurred on the lower third of the pipe, in the water phase. Preferential corrosion of carbon steel weld metal occurs in many different corrosive media. Its precise cause is still under study, but there are general indications that silicon additions and basic coated welding consumables accelerate the effect. Low levels of nickel, chromium and copper generally are beneficial.
Remedy: Use welding consumables with minimum silicon content and additions of up to 0.5% Ni, Cu or Cr.

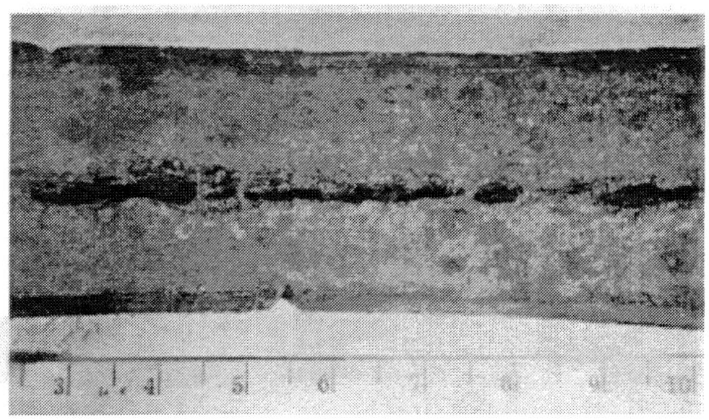

Figure 2.5

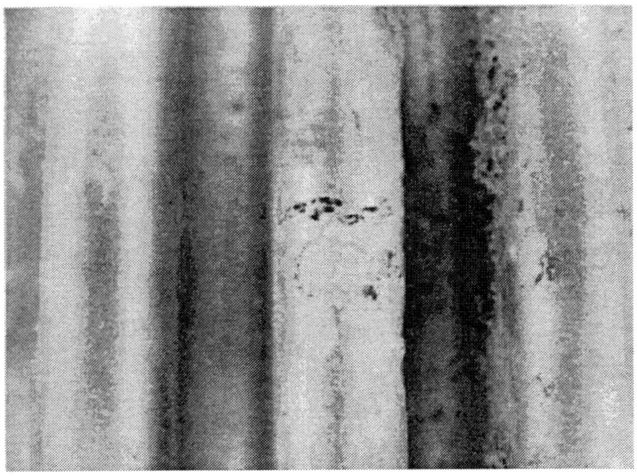

Figure 2.6 - macro

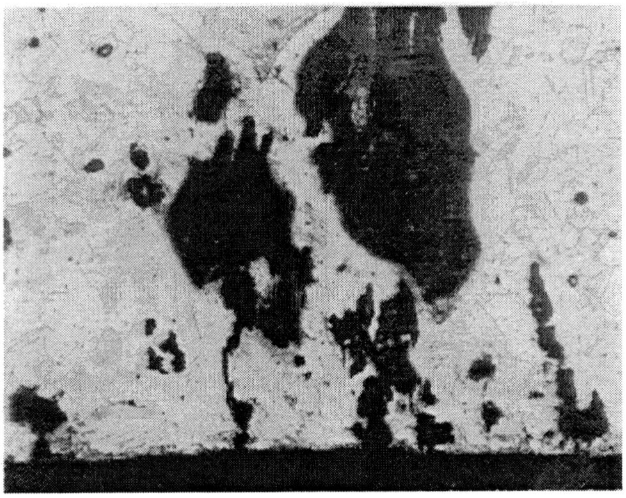

Figure 2.7(a) - 150X

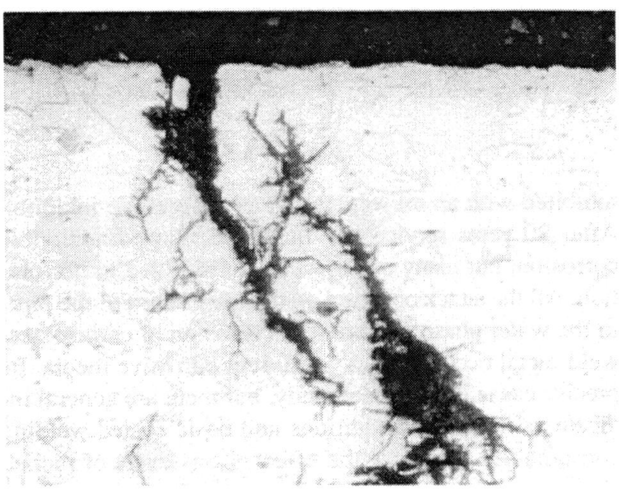

Figure 2.7(b) - 150X

CASE HISTORY 2.6

Phenomenon: pitting
Group/ Type: AISI 316 SS (UNS S31600)
Equipment: product transfer hose
Environment: marine atmosphere
Conditions:
(a) major species: air, rain water
(b) minor species: sand, chlorides
(c) temperature: ambient
(d) pressure: atmospheric
(e) aeration: saturated
Time to failure: 1 year
Illustration: macro
Comments: Eleven sections of AISI 316 SS (UNS S31600) braided, corrugated metal transfer hose were stored outdoors on the ground at a marine terminal on Galveston Bay for one year. When first used after this time, all 11 sections were found to be perforated by classic chloride pitting of the heavily cold-worked, corrugated metal. In addition, some of the reinforcing braid that had rested directly on the sand was corroded completely.
Remedy: Store stainless transfer hose under cover, on raised racks, with end caps to prevent the accumulation of rainwater or runoff.

CASE HISTORY 2.7

Phenomenon: pitting and SCC
Group/ Type: AISI 304 SS (UNS S30400)
Equipment: corrosion coupon
Environment: ferric sulfate solution
Conditions:
(a) major specimens: water, 20% ferric sulfate
(b) minor species: 20 ppm chloride ion
(c) temperature: 140°F (60°C)
(d) pressure: 1 atm
(e) aeration: natural, without aeration
Time to failure: 14 days
Illustration: micros
Comment: The iron ions in ferric sulfate solutions, together with their low pH, make potent pitting and cracking environments for stainless steels if chlorides are present. Both pitting and stress corrosion cracking were found on this supposedly unstressed corrosion coupon in a laboratory beaker test.
Remedy: Substitute high-molybdenum ferritic SS or chromium-nickel-molybdenum alloys.

CASE HISTORY 2.8

Phenomenon: pitting
Group/ Type: AISI 304 SS (UNS S30400)
Equipment: 4-in., schedule 10 piping
Environment: city water
Conditions:
(a) major species: water
(b) minor species: 150 ppm Cl-
(c) temperature: ambient
(d) pressure: atmospheric
(e) aeration: variable
Time to failure: 18 months
Illustration: macro
Comment: After hydrotest, city water was allowed to stand in a run of AISI 304 SS (UNS S30400) piping. Approximately 18 months after initial hydrotest, the line leaked on startup. Examination revealed extensive pitting in areas where the water had dried, leaving patches of concentrated water salts. Reddish-brown to black stains signaled locations where pitting was most severe.
Remedy: Replace in kind, dry, and cap after hydrotest.

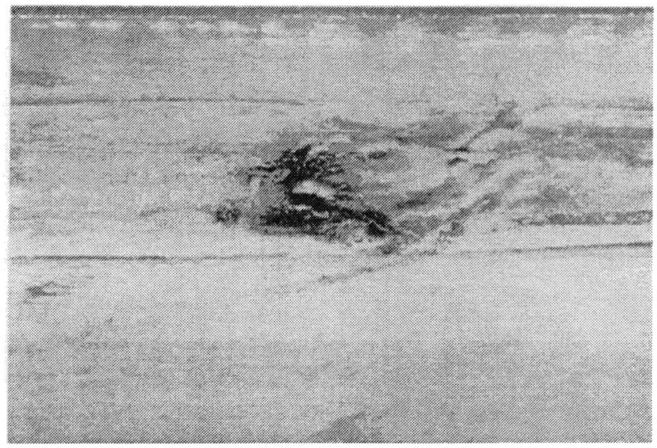

Figure 2.8 - macro

CASE HISTORY 2.9

Phenomenon: pitting, followed by SCC
Group/ Type: ferrous, pH 13-8Mo SS, condition H1000 (UNS S13800)
Equipment: laboratory bent beam specimen
Environment: acidified seawater
Conditions:
(a) major species: water, 3½% NaCl, 0.5% acetic acid
(b) minor species:
(c) temperature: 100°F (37.8°C)
(d) pressure: atmospheric
(e) aeration: saturated
Time to failure: 40 days
Illustration: macro, SEM fractograph
Comment: Precipitation-hardening SS in the fully overaged conditions have reasonably good SCC resistance. However, as the hardness goes up, the SCC resistance goes down. This specimen, in the slightly overaged H1000 condition, did not fail in the fully exposed center of the beam. However, a silicone rubber sealant had been used to mask off the steel mounting screws and the matching holes in the beam. The silicone rubber pulled away from the specimen with time, creating a crevice where locally high acidity and chloride levels allowed crack initiation.
Remedy: Use the alloy in the fully overaged condition for chloride service.

Figure 2.9(a) - 45X

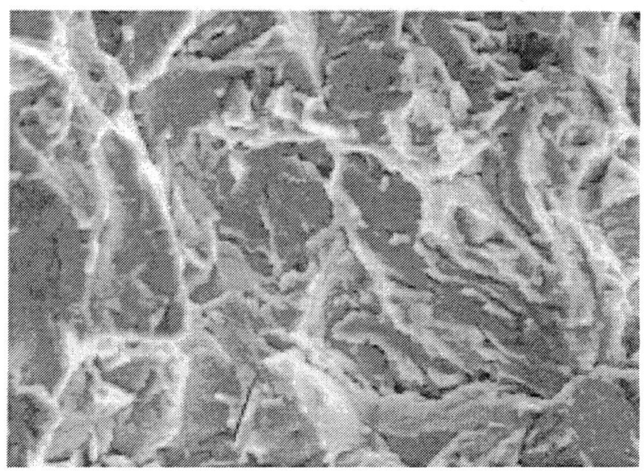

Figure 2.9(b) - 1,000X

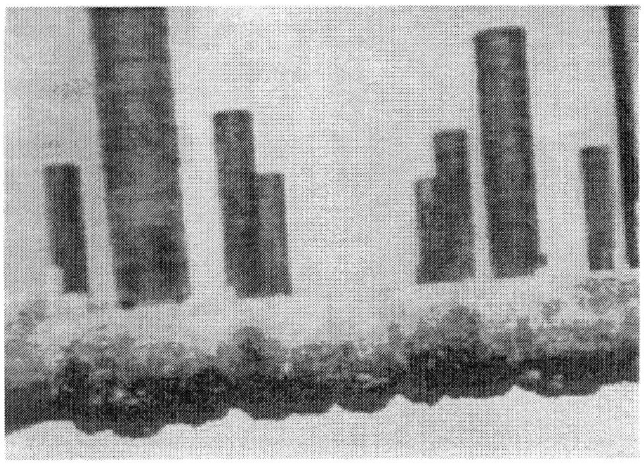

CASE HISTORY 2.10

Phenomenon: pitting caused by microbial activity
Group/ Type: AISI 304 SS (UNS S30400)
Equipment: water treatment tank
Environment: waste water
Conditions:
(a) major species: waste water, bulk pH 7
(b) minor species: air, CO_2, NH_3, nitrates
(c) temperature: Texas Gulf Coast ambient
(d) pressure: 1 atm
(e) aeration: forced
Time to failure: 6 months
Illustration: macro with deposits, after cleaning

Comment: This open-topped tank treated waste water by microbial oxidation. After six months, the forced air piping was covered on the stagnant bottom side by a thick layer of sludge with voluminous slime deposits. Cleaning revealed broad, round pits with terraced sidewalls. The piping was perforated at many pits. Analysis of the slime showed a pH of 3 to 4, with high counts of both sulfate-reducing and iron bacteria.
Remedy: Sandblast, weld repair, and coat with epoxy. Place the vessel under cathodic protection.
Reference: G. Kobrin, "Reflections on Microbiologically Induced Corrosion of Stainless Steels," E.I. Dupont de Nemours, Inc.

CASE HISTORY 2.11

Phenomenon: crevice corrosion caused by microbial activity
Group/ Type: AISI 304 SS (UNS S30400)
Equipment: cooling tower piping
Environment: Cape Fear, NC river water
Conditions:
(a) major species: water, silt
(b) minor species: iron ions, 0.5-1.0 ppm chlorine residual, biocide (quaternary amine + tris tri-butyltin oxide)
(c) temperature: 104°F (40°C)
(d) pressure: not given
(e) aeration: saturated
Time to failure: 3 years
Illustration: macro of deposits, macro after cleaning (page 25)

Comment: Severe crevice corrosion occurred in and near gasketed flanged joints, in low velocity areas, under voluminous tannish-brown slime deposits. The tan deposits were high in slime and iron bacteria. Nearby black deposits were high in sulfate-reducing bacteria. After cleaning, these crevice sites showed broad, open pits with bright, active surfaces. Slime-forming bacteria apparently trapped iron rich solids and promoted iron bacteria growth. These aerobic bacteria consumed available O_2, producing anaerobic conditions under the deposits that were ripe for sulfate-reducing bacteria. Oxygen depletion and sulfate reduction caused crevice corrosion on the stainless, releasing iron ions, which helped perpetuate the cycle.
Remedy: Repair flange faces with epoxy coating, use non-wicking rubber gaskets instead of asbestos.
Reference: G. Kobrin, "Reflections on Biologically Influenced Corrosion of Stainless Steels," E.I. DuPont de Nemours & Co.

CASE HISTORY 2.12

Phenomenon:
pitting, followed by SCC, corrosion fatigue, or both.
Group/Type: AISI 316L SS (UNS S31600)
Equipment:
Kuntscher nails (femoral orthopedic implants)
Environment: body fluids, bone marrow

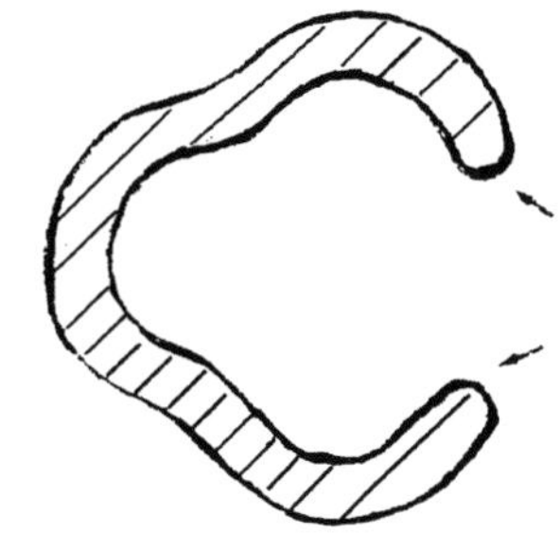

Conditions:
- (a) major species: 0.9% total salts in water
- (b) minor species: Ca, P, S, organics, (pH = 7.4)
- (c) temperature: 99°F (37°C)
- (d) pressure: atmospheric
- (e) aeration: 4 ppm O_2 (65% of saturation)

Time to failure: less than 12 months
Illustration: cross section
Comment: The low levels of chlorides in body fluids are enough to initiate pitting at crevices in standard austenitic SSs. Cracking failures can initiate at the stress raisers formed by the pits. An estimated 3% of implanted nails must be replaced prematurely because of such failures. This particular fracture showed characteristics of both SCC and corrosion fatigue. The cyclic stresses of normal walking are high enough that a cyclic component to the crack growth is plausible.
Remedy: Replace with cobalt-based or titanium-based alloys.
Reference: G. Bombara, M. Cavallini, *Corrosion Science*, 17 (1977): pp. 77-85. (Drawing reprinted with kind permission from Elsevier Science Ltd., The Boulevard, Langford Lane, Kidlington OX5 1GB, UK.)

Figure 2.13 - Pitted inner surface of pipe.

CASE HISTORY 2.13

Phenomenon: pitting corrosion
Group/ Type: ferrous, AISI type 316 SS
Equipment: piping, seawater treatment plant
Environment: seawater plus sulfur dioxide
Conditions:
- a) major species: water, 5% NaCl
- b) minor species: 0.5% SO_2
- c) temperature: 100ºF (37.8°C)
- d) pressure: atmospheric
- e) aeration: deaerated

Time to failure: 1 month
Illustration: photograph
Comment: Sulfur dioxide from a sulfur burner was bubbled through a seawater extractor to make a deaerated high-chloride aqueous solution with an estimated pH of 1.2. A 1½-inch diameter piping and metering system for this unit was inadvertantly made out of AISI 316 stainless steel. The piping perforated due to pitting corrosion within one month of unit startup.
Remedy: Replace with non-metallic piping (CPVC or RTR) with orifice plates and valves of high-molybdenum, high-chromium nickel-based alloys.

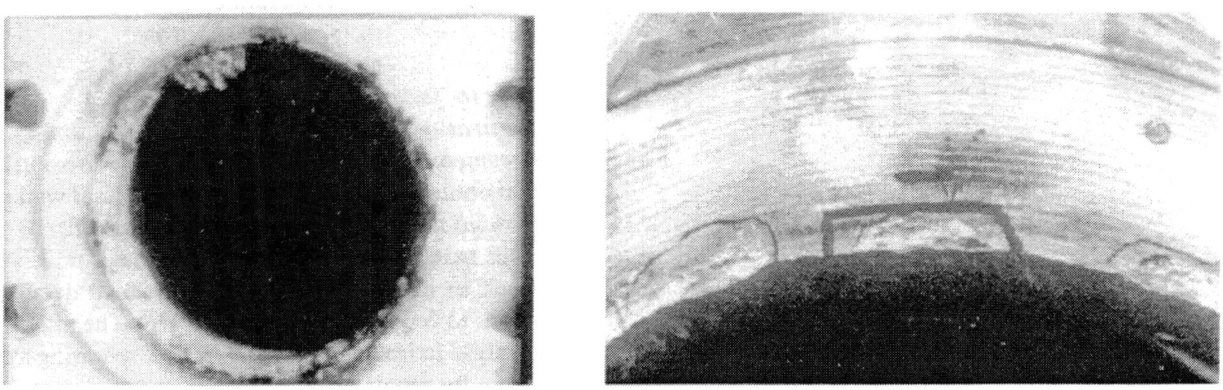

Figure 2.11 - (a) Macro of deposits (b) and macro after cleaning.

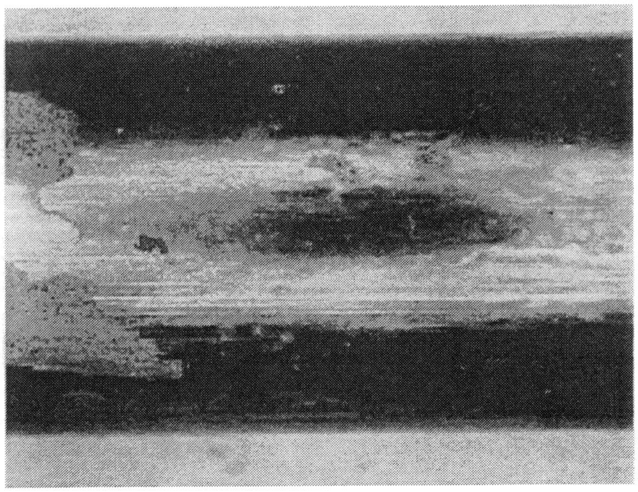

Figure 2.14 - Macro

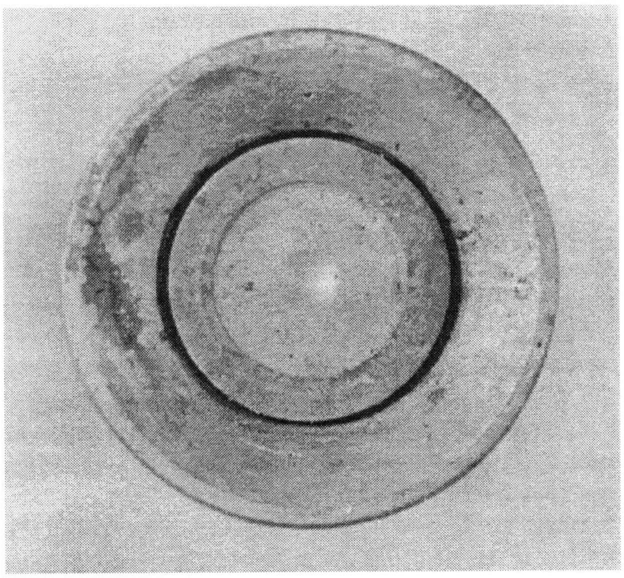

Figure 2.15 - Photomacrograph

CASE HISTORY 2.14

Phenomenon: pitting, spalling
Group/ Type: ferrous, 17-4 PH stainless (condition H1100) (UNS S17400) chrome oxide coating
Equipment: steam pump rod
Environment: process fluid
Conditions:

(a) major species: steam, vinyl chloride, acetone
(b) minor species: HCl
(c) temperature: 257°F (125°C)
(d) pressure: not recorded, flow rate 250 gal/min.
(e) aeration: none

Time to failure: 5 months
Illustration: macro
Comments: This 17-4PH SS piston rod was coated with a chromium-oxide coating to improve the life of mechanical packing. The chromium-oxide coating had an epoxy binder, which was quickly attacked by the process. Once the binder had been dissolved, pitting of the 17-4PH stainless by traces of HCl in the process led to spalling, blistering, and flaking of the oxide coating.
Remedy: Use a tungsten carbide coating with cobalt binder, or convert to alloy C-276 rods.

CASE HISTORY 2.15

Phenomenon: pitting
Group/ Type: AISI 316 SS (UNS S31600)
Equipment: safety valve and valve seat
Environment: isopropylchloroformate
Conditions:

(a) major species: isopropylchloroformate, steam
(b) minor species: HCl, HF
(c) temperature: 257°F (125°C)
(d) pressure: 135 psig
(e) aeration: none

Time to failure: 6 days
Illustration: photomacrograph
Comments: Relatively minor general corrosion or pitting that could be tolerated in a pipe or vessel wall may lead to rapid failure of items requiring a good fit to function. This safety valve and valve seat pitted in less than one week in isopropylchloroformate, which decomposes in steam to form traces of HCl and HF. The pitting allowed catalyst to leak past the safety valve, requiring a unit shutdown for repair.
Remedy: Replace with nickel-copper alloy 400 valve trim.

CASE HISTORY 2.16

Phenomenon: pitting
Group/ Type: AISI 316L SS (UNS S31603)
Equipment: storage tank
Environment: merchant grade wet process phosphoric acid
Conditions:
(a) major species: 53% H_3PO_4, H_2O
(b) minor species: traces of NaCl
(c) temperature: 122°F (50°C) maximum
(d) pressure: atmospheric
(e) aeration: saturated by agitation
Time to failure: 2 years
Illustration: macro
Comment: These storage tanks were of steel clad with AISI 316L SS. Great care was taken to ensure that chloride levels remained low by freshwater washing and clean room entry procedures. The condition of the cladding after ~ two years of service was generally good. However, some pitting was observed at areas of weld spatter and arc strikes, where the defect shape promoted crevice attack and the lack of shielding led to contamination of the molten metal with oxygen and nitrogen.
Remedy: Remove weld spatter by grinding after welding. Follow with a nitric acid passivation treatment.

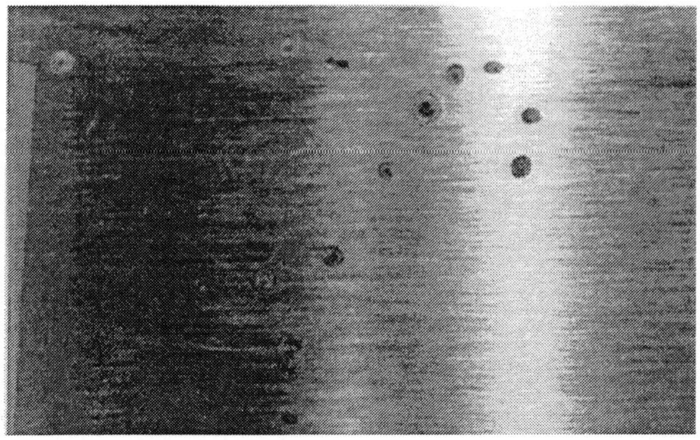

Figure 2.16 - macro

CASE HISTORY 2.17

Phenomenon: crevice corrosion
Group/Type: AISI 304L SS (UNS S30403) with 308L weld
Equipment: refrigeration coils
Environment: brine
Conditions:
(a) major species: water, 20% NaCl
(b) minor species:
(c) temperature: 7°F (-13.9°C)
(d) pressure: not given
(e) aeration: not determined
Time to failure: 3 years
Illustration: macro
Comment: An epoxy coating was applied over these welds after fabrication. Lack of adhesion allowed brine to seep underneath the coating, where conditions for severe crevice corrosion were established.
Remedy: Do not coat crevice-sensitive materials.
Contributor: A.C. Holler, Twin City Testing and Engineering Laboratory, Inc., St. Paul, Minn.

Figure 2.17 - macro

CASE HISTORY 2.18

Phenomenon: pitting and galvanic attack
Group/ Type: AISI 304 SS (UNS S30400)
Equipment: activated carbon absorber tank
Environment: salt water + dissolved organic
Conditions:

(a) major species: 5% NaCl in water, pH 3 to 4
(b) minor species: dissolved organic
(c) temperature: 68°F (20°C), steady state (steamed out weekly to reactivate the carbon particles)
(d) pressure: not given
(e) aeration: not determined

Time to failure: 1 month
Illustration: potentiodynamic polarization scan
Comment: The corrosion potential of the AISI 304 SS (UNS S30400) tank in this 5% NaCl stream was shifted from its normal value of ~ -180 mV vs Ag/AgCl to an estimated +120 mV vs Ag/AgCl when coupled to the carbon particles in this activated carbon absorber. This potential shift thrust the SS squarely into the midst of the transpassive zone, where pitting at catastrophic rates can occur (Figure). The resulting perforation of the 0.188-in. thick tank wall at numerous pits in one month means an average pitting rate of 2.25 in./year.
Remedy: Use laboratory observations of galvanic current density to estimate pitting rate of candidate replacement materials, assuming that one micro-amp/cm² = 0.010mm/y general corrosion rate, concentrated in 10% of the specimen's exposed area.
Reference: E.L. Liening "Trouble-shooting Industrial Corrosion Problems with Electrochemical Testing Techniques", ASTM Symposium on Non-destructive Testing and Electrochemical Methods for Monitoring Corrosion in Industrial Plants, ASTM, Philadelphia, PA, 1984.

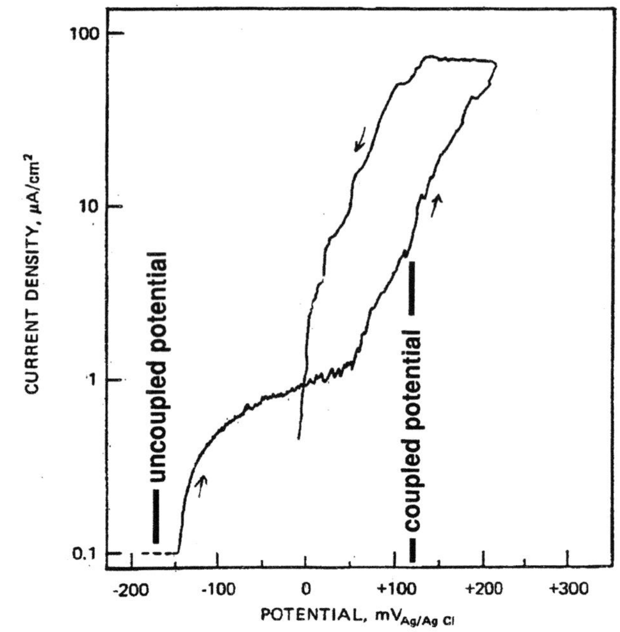

CASE HISTORY 2.19

Phenomenon: preferential phase attack in welds
Group/ Type: AISI 304L SS (UNS S30403)
Equipment: nitrator coil
Environment: nitroglycerin manufacture
Conditions:

(a) major species: HNO_3, H_2SO_4
(b) minor species: not given
(c) temperature: 61°F (16°C)
(d) pressure: not given
(e) aeration: not given

Time to failure: not given
Illustration: micro
Comment: Ferrite suffers preferential corrosion in some mixtures of nitric and sulfuric acid. If the ferrite pools in austenitic-ferritic SS welds are interconnected, as in this case, preferential corrosion can cause a path of leakage through the welds.
Remedy: Weld with a fully austenitic filler metal.
Reference: K.F. Krysiak, "Cause and Prevention of Unusual Failures of Materials," CORROSION/83, paper no. 19 (Houston, TX: NACE, 1983).

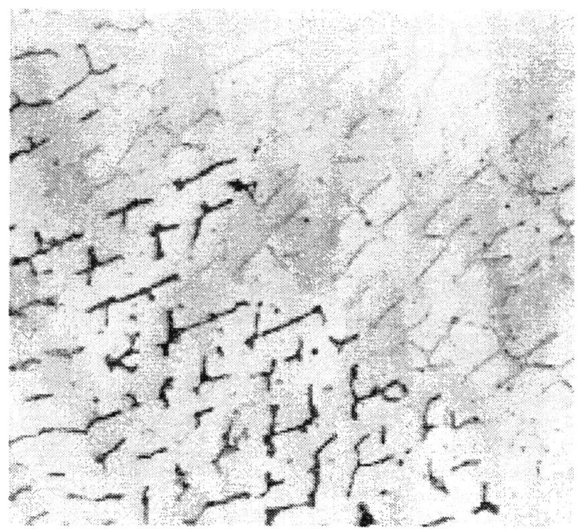

Figure 2.19 - micro

CASE HISTORY 2.20

Phenomenon: pitting
Group/ Type: AISI 430 SS (UNS S43000)
Equipment: heat exchanger tubes
Environment: refinery sour water
Conditions:
(a) major species: water, 5,000 ppm H_2S, pH 7
(b) minor species: 90 ppm mercaptans, 10 to 30 ppm Cl-
(c) temperature: 86°F (30°C) inlet, 212°F (100°C) outlet
(d) pressure: 145 psig
(e) aeration: not given
Time to failure: 6 months
Illustration: schematic
Comment: Sulfide and chloride combinations make powerful pitting and cracking environments for many common SS. This feed/effluent heat exchanger suffered rapid pitting failures of AISI 430 (UNS S43000) stainless tubes. Replacement tubes of AISI 316 SS also failed because of SCC within 6 months.
Remedy: UNS S31500 (18 Cr, 5 Ni, 3 Mo) tubes have been in service nine years with no failure
Reference: C. Martensen, G. Supko, "Experience with High-Alloy Stainless Steels in the HPI," *Hydrocarbon Processing*, 1 (1983): pp. 69-74.

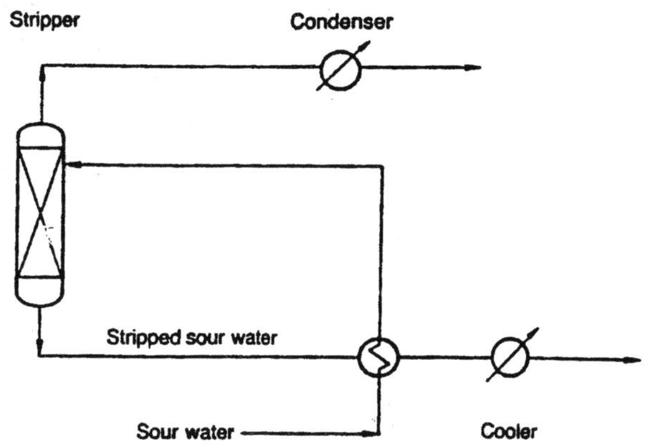

Figure 2.20 - Schematic.

CASE HISTORY 2.21

Phenomenon: crevice corrosion
Group/ Type: SS, ferritic grade 29-4-2 (UNS S44800)
Equipment: heat exchanger tubing
Environment: flue gas desulfurizing process
Conditions:
(a) major species: flue gas from coal with 2.9% S
(b) minor species: H_2SO_4 - 12 ppm, 5% H_2O
(c) temperature: 174 to 203°F (79 to 95°C)
(d) pressure: 1 atm
(e) aeration: not given
Time to failure: 3,060 hours
Illustration: macro, micro
Comment: Small water leaks around the sealing mechanism of these test heat exchanger tubes caused local dilution of condensed sulfuric acid to particularly damaging concentrations.
Remedy: Seal weld
Reference: H.H. Krause, P.A. Ireland, G.A. Charles, R.G. Rhudy, B.C. Syrett, "Cyclic Reheat Materials Evaluation," *Proc. CORROSION/84 Symposium on Corrosion in Flue Gas Desulfurization Systems* (Houston, TX: NACE 1984): pp. 353-370.

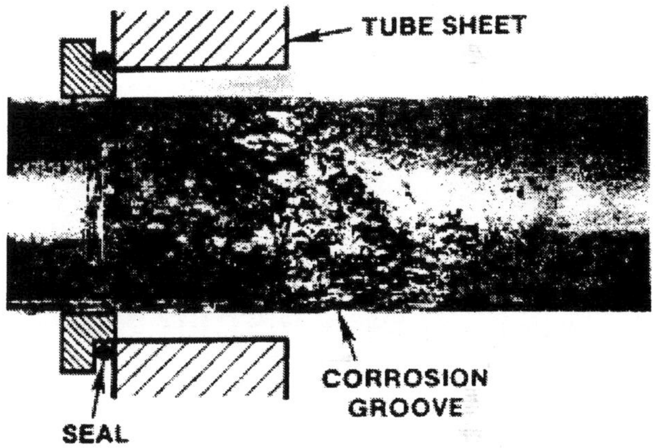

Figure 2.21(a)

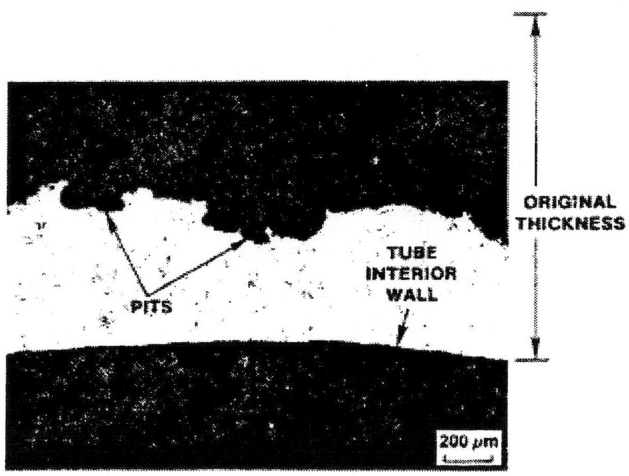

Figure 2.21(b)

CASE HISTORY 2.22

Phenomenon: pitting corrosion
Group/ Type: ferrous, AISI 317L (UNS S31703)
Equipment: air-cooled heat exchanger
Environment: acetic acid production
Conditions:
(a) major species: 80 to 85% ace tic acid
(b) minor species: 5 to 10% for mic acid,
5 to 12% water
(c) temperature: 248°F (120°C) inlet, 104°F (40°C) outlet
(d) pressure: 870 psig
(e) aeration: not given
Time to failure: 1 year
Illustration: schematic
Remedy: UNS N08904 (20Cr, 25Ni, 4.5Mo) tubes have been in service for six years with no failure.
Reference: C. Martenson, G. Supro, "Experience with High-Alloy Stainless Steels in the HPI," *Hydrocarbon Processing,* (1983): pp 69-74.

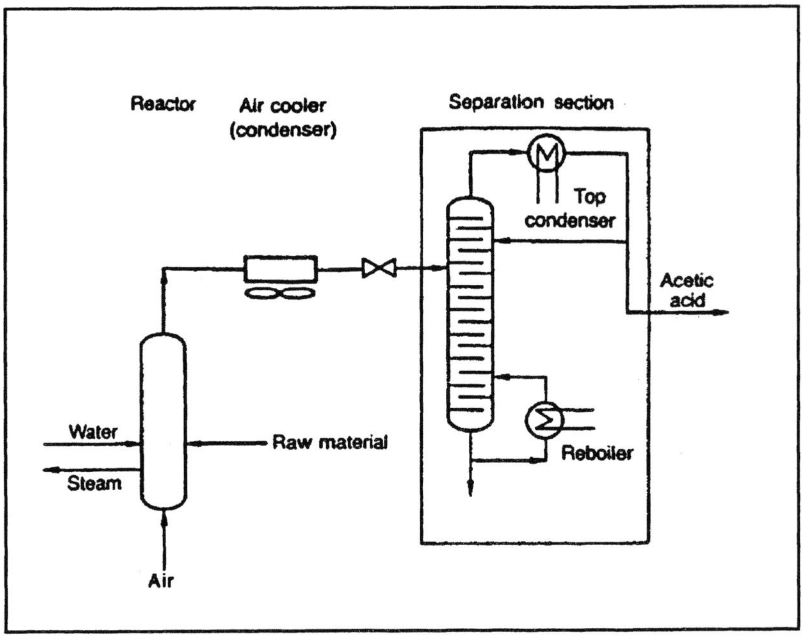

CASE HISTORY 2.23

Phenomenon: pitting
Group/ Type: AISI type 304 SS (UNS S30400)
Equipment: strip lining in a deethanizer reboiler shell
Environment: sour natural gas liquids
Conditions:
a) major species: methane, ethane and propane
b) minor species: 5 wt% H_2S, traces of water
c) temperature: 250ºF (121°C)
d) pressure: 410 psia
e) aeration: anaerobic
Time to failure: 10 years
Illustration: macro
Comment: When austenitic stainless steels pit, it is often assumed that chlorides must be the culprit. However, the complex reducing sulfurous acids that develop in wet sour NGL can also serve as pitting agents on stainless steels. After 10 years in sour NGL service, a band of broad, shallow pits developed along the bottom of this horizontal AISI 304 SS strip lined steel vessel. This service is hot and high in hydrogen sulfide, but free of chlorides.
Remedy: Replace the AISI 304 SS strip lining with AISI 316L SS.

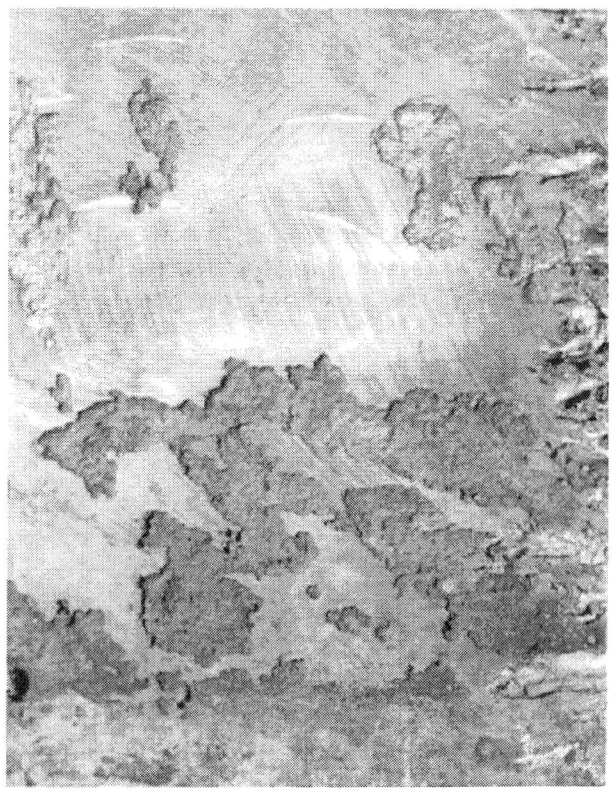

Figure 2.23 - Pitting of welded strip liner.

CASE HISTORY 2.24

Phenomenon: pitting
Group/ Type: AISI 304 SS (UNS S30400)
Equipment: salt evaporator
Environment: brine
Conditions:
(a) major species: water, NaCl
(b) minor species: occasional air leakage
(c) temperature: not given
(d) pressure: not given
(e) aeration: erratic, with freshwater injection or loss of vacuum seal

Time to failure: 3 days
Illustration: drawing, potentiodynamic polarization curve
Comment: A SS salt evaporator with cast iron inlet piping operated successfully for many years in a European chemical plant. An "identical" unit, but one with all SS piping, was built in North America. The new unit failed in three days. Investigation revealed that the natural potential of SS in this process was in the pitting range (Figure). The European unit was receiving just enough protective cathodic current from its cast iron piping to eliminate pitting of the SS. The "improved" North American unit, with its all SS piping, had no such protection.
Remedy: Installation of a controlled-output, galvanic-anode cathodic protection system successfully eliminated further pitting on the SS tubes.
Reference: H.A. Webster, "Innovation in the Practical Application of Electrochemical Technology in Corrosion Control," presented at the NACE Eastern Canadian Regional Meeting, 1982.

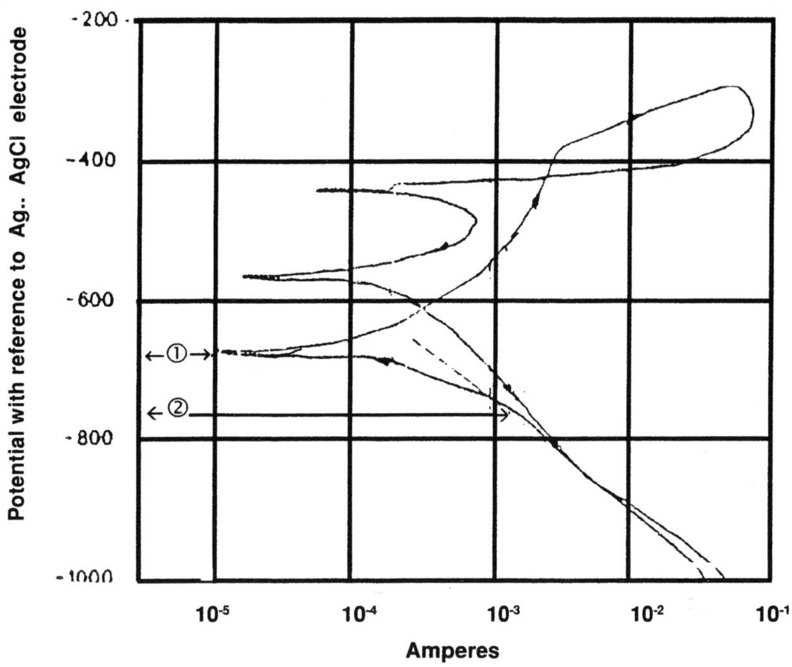

1. open circuit potential 2. coupled to cast iron

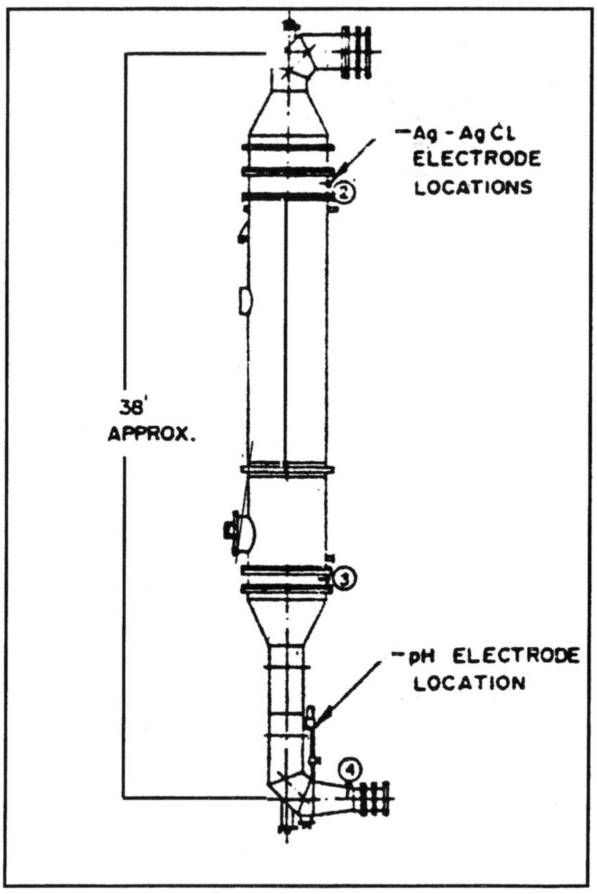

Figure 2.24(b) - (right) Stainless steel circulating pipework - salt evaporating system.

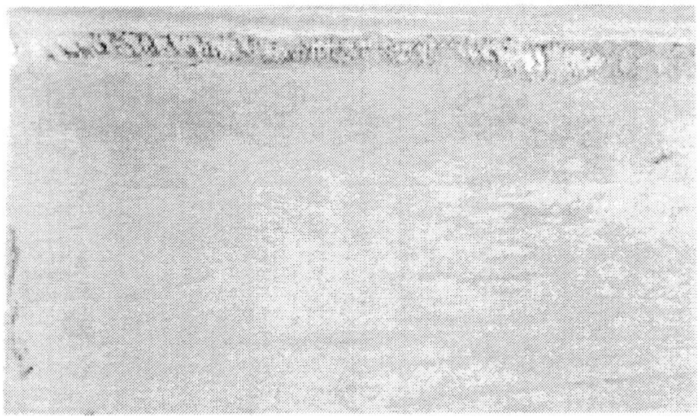

Figure 2.25 - Macro

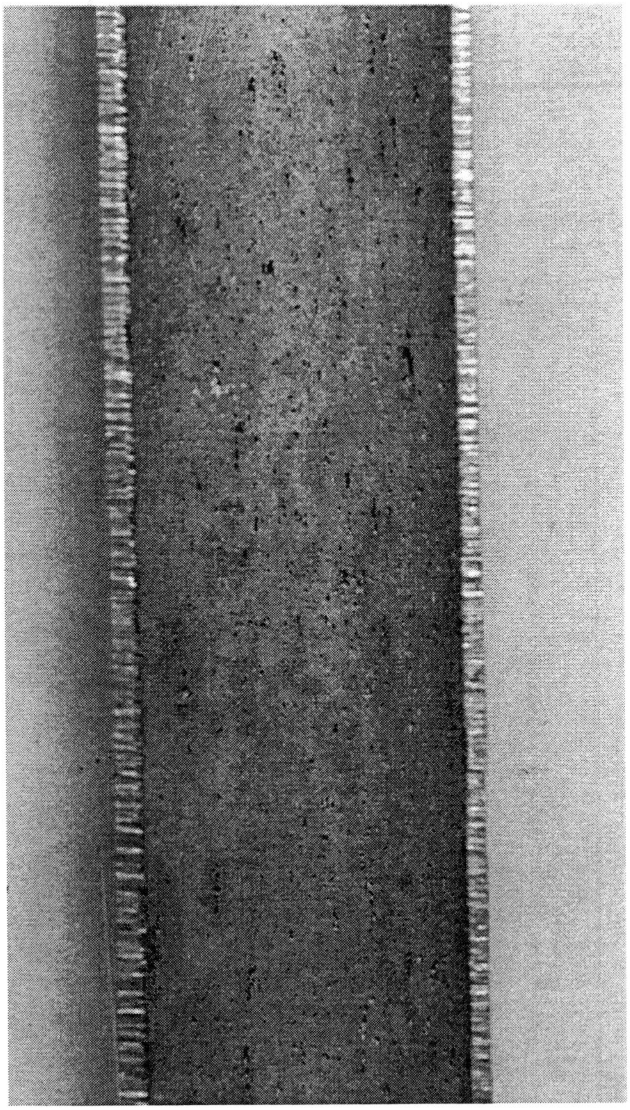

Figure 2.26 - Macro

CASE HISTORY 2.25

Phenomenon: selective weld metal attack
Group/Type: nickel-based alloy: alloy C-276 (UNS N10276)
Equipment: deflector plate
Environment: hydrofluoric acid
Conditions:
(a) major species: 50% HF
(b) minor species: water
(c) temperature: atmospheric boiling
(d) pressure: atmospheric (flow rate: 50 ft/s)
(e) aeration: none

Time to failure: 4 days
Illustration: macro
Comments: In most environments, alloy C-276 reportedly shows no difference between welds and base plate with regard to corrosion performance. However, the high-velocity hydrofluoric acid in this case caused selective attack of the welds to depths on the order of 0.02 in. in only 4 days, corresponding to a corrosion rate of 1.8 in./y in the weld metal.
Remedy: Use alloy C-22 for this service.

CASE HISTORY 2.26

Phenomenon: pitting
Group/Type: nickel-based alloy (UNS N06600)
Equipment: heat exchanger tubes
Environment: paper mill digester liquor
Conditions:
(a) major species: H_2O, NaOH, sulfates, cellulose
(b) minor species: polythionates, HCl (intermittent)
(c) temperature: 374°F (190°C) (steady state), 122°F (50°C) during cleaning
(d) pressure: not given
(e) aeration: not given

Time to failure: 4 years
Illustration: macro
Comment: Alloy 600 (UNS N06600) has good resistance to the caustic-sulfate environments in paper mill digesters. However, this heat exchanger was periodically descaled with HCl. Alloy 600 (UNS N06600) with no molybdenum and a relatively low chromium content has poor pitting resistance in acid chloride solutions.
Remedy: Descale mechanically or with sulfamic acid.

CASE HISTORY 2.27

Phenomenon: pitting corrosion

Group /Type: nickel-iron-chromium alloy, UNS N08800

Equipment: expansion joint bellows in fluid catalytic cracker regenerator standpipe

Environment: steam, torch oil combustion products

Conditions:

a) major species: regenerated alumina-silica catalyst particles
b) minor species: condensed steam, SO_2, H_2S
c) temperature: 138ºC (280°F)
d) pressure: 40 psig
e) aeration: erratic

Time to failure: 5 months

Illustration: macro, SEM micro, energy dispersive x-ray spectrum

Comment: This alloy 800H bellows was given a 1,700°F solution treatment to maximize its creep resistance. The bellows was then installed in a cold-wall expansion joint on the regenerated catalyst standpipe of a fluidized catalytic cracker (FCC). When first installed this unit suffered an extended startup due to a slide valve malfunction just downstream of the expansion joint. During this period steam and sulfur-bearing torch oil combustion products apparently combined to condense in the convolutions of the bellows, creating pools of sulfurous polythionic acids. These acids proved to be potent pitting agents on the sensitized alloy 800H bellows.

Remedy: Replace with an alloy 625 (UNS N06625) bellows of two-ply design with a pressure port between plies to signal failure of the inner ply.

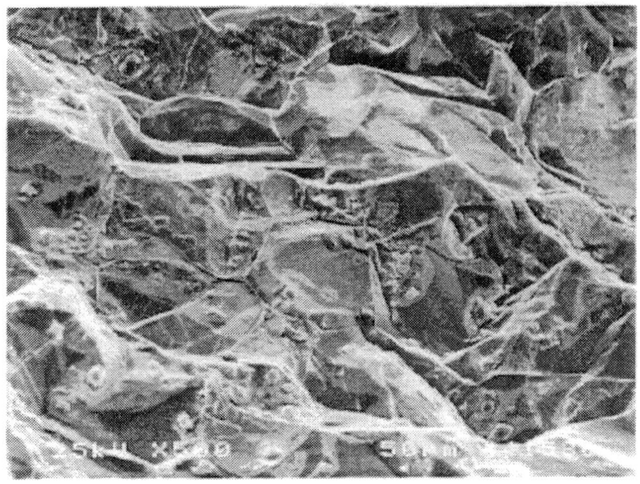

CH 2.27(a) - SEM micrograph of pit in bellows

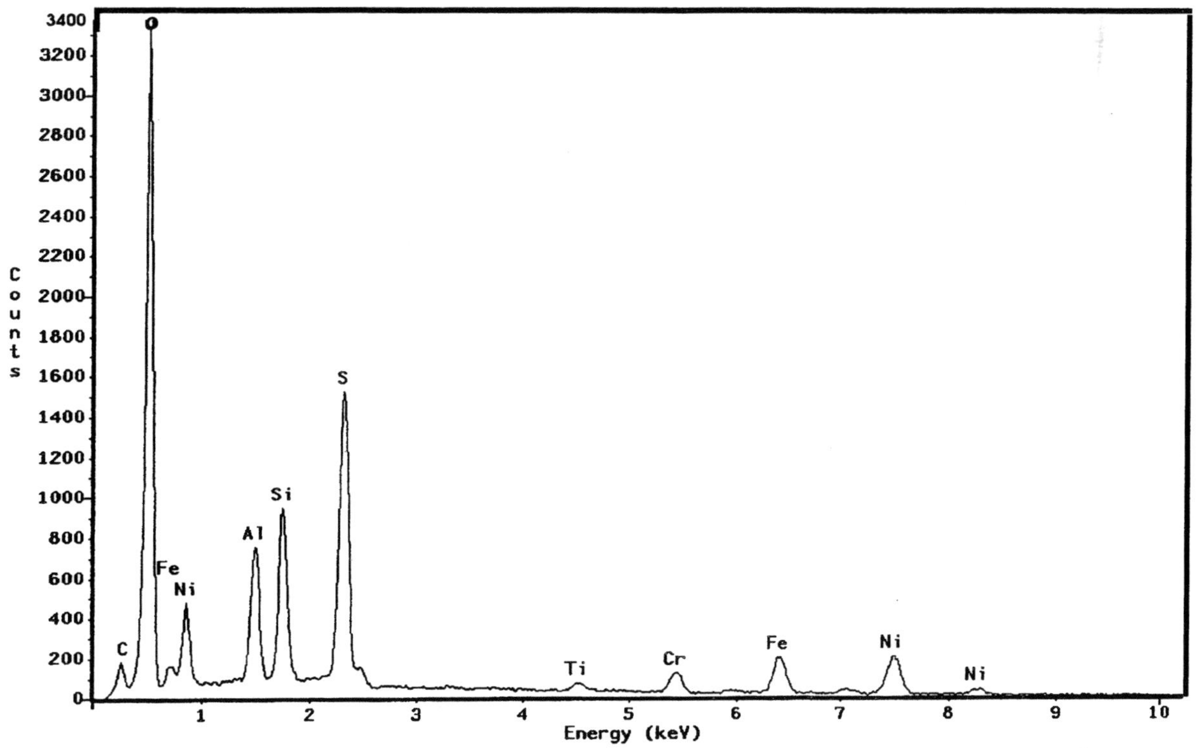

CH 2.27(b) - Energy dispersive x-ray analysis of corrosion products in pitted bellows.

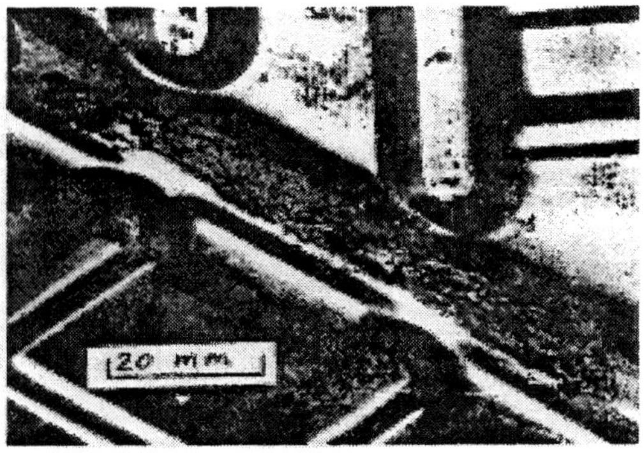

Figure 2.28(a) - EDX spectra

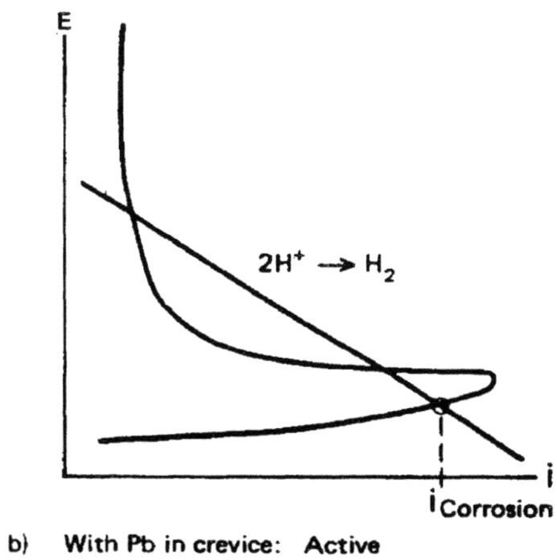

Figure 2.28(b) - Anodic polarization curve

CASE HISTORY 2.28

Phenomenon: crevice corrosion
Group/ Type: titanium, commercially pure grade 2 (UNS R50400)
Equipment: plate-and-frame heat exchanger
Environment: 26% NaCl brine
Conditions:
(a) major species: NaCl, H_2O, pH 8
(b) minor species: not determined
(c) temperature: inlet, ambient; outlet, 167°F (75°C)
(d) pressure: not specified
(e) aeration: not determined
Time to failure: 30 days
Illustration: EDX spectra, anodic polarization curve
Comment: A crevice corrosion rate of 330 mpy on titanium heat exchanger plates was attributed to the use of fluorinated rubber gaskets containing lead oxide curing agent. The lead oxide was apparently leached from the rubber then reacted with the brine to form $PbCl_2$ particles. Apparently, the lead, which has a very low exchange current density, interfered with the cathodic reaction to the point the cathodic current was reduced below the level needed to overcome the anodic nose of the current-voltage curve (see Figure); unstable surface passivity resulted.
Remedy: A replacement titanium exchanger with EPDM rubber gaskets had lasted 12 months with no failure at the time of publication.
Reference: E.L. Liening, *Materials Performance* 11(1983): pp 37-44.

CASE HISTORY 2.29

Phenomenon: pitting
Group/ Type: titanium alloy, grade 12 Ti+Ni+Mo
Equipment: heat exchanger tubes
Environment: well brine
Conditions:
(a) major species: 20% $CaCl_2$ in well water
(b) minor species: K, Br, and I ions
(c) temperature: 203°F (95°C)
(d) pressure: not specified
(e) aeration: not determined
Time to failure: 60 days
Illustration: process schematic, pit schematic (right)
Comment: Although grade 12 titanium alloy had demonstrated superior corrosion resistance to this environment in laboratory tests, 86 out of 563 tubes in this heat exchanger began leaking within 60 days of startup. Although the failure analysis was inconclusive, the most plausible explanation was believed to be that iron particles rolled into the tube surfaces corroded to form occluded cells with locally high iron ion content and low pH. The acidity and high iron ion content prevented surface passivation of the titanium beneath the particles, leading to rapid perforation by pitting.
Remedy: Pickle heat exchanger tubes to remove Fe after fabrication.
Reference: E.L. Liening, "Unusual Failures of Titanium Equipment in Chemical Processing," CORROSION/83, paper 15 (NACE, 1983).

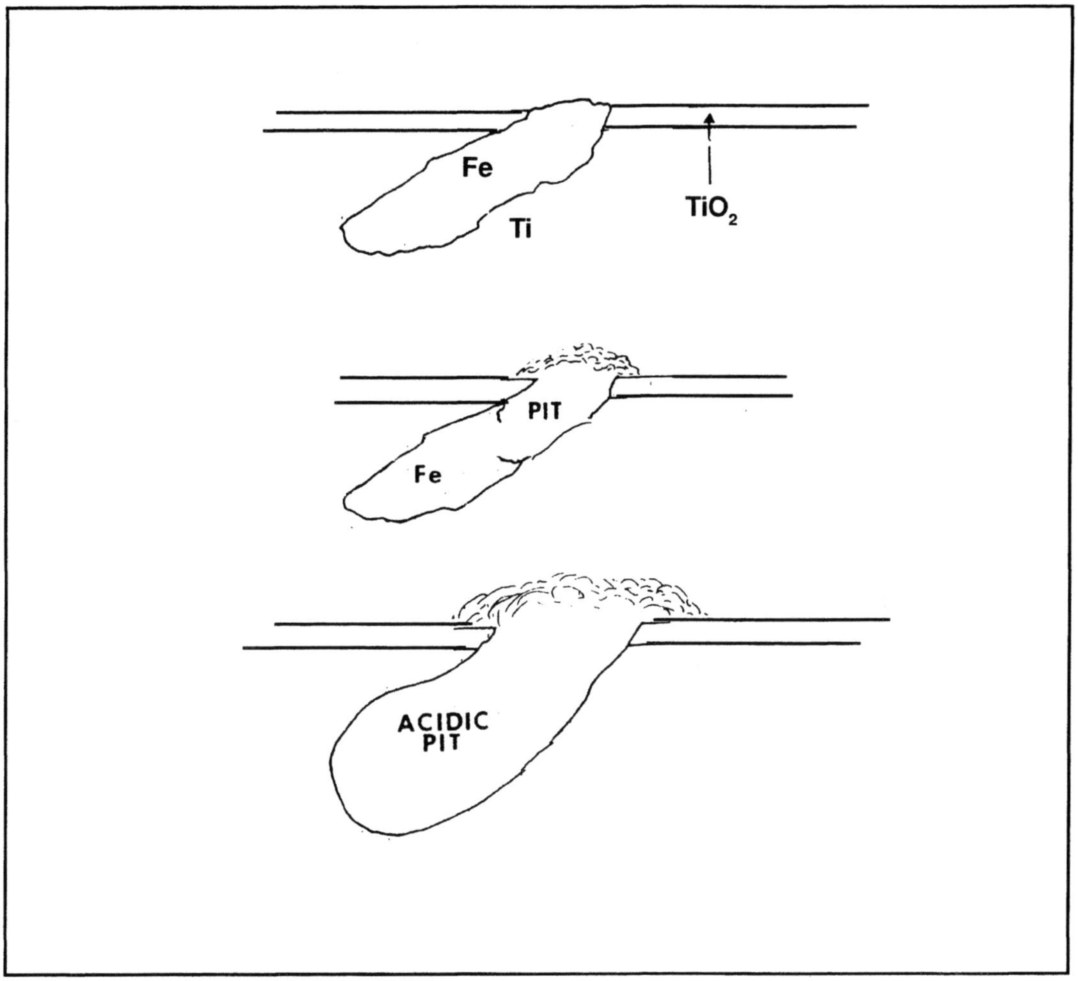

Figure 2.29 - (a) pit schematic and (b) process schematic

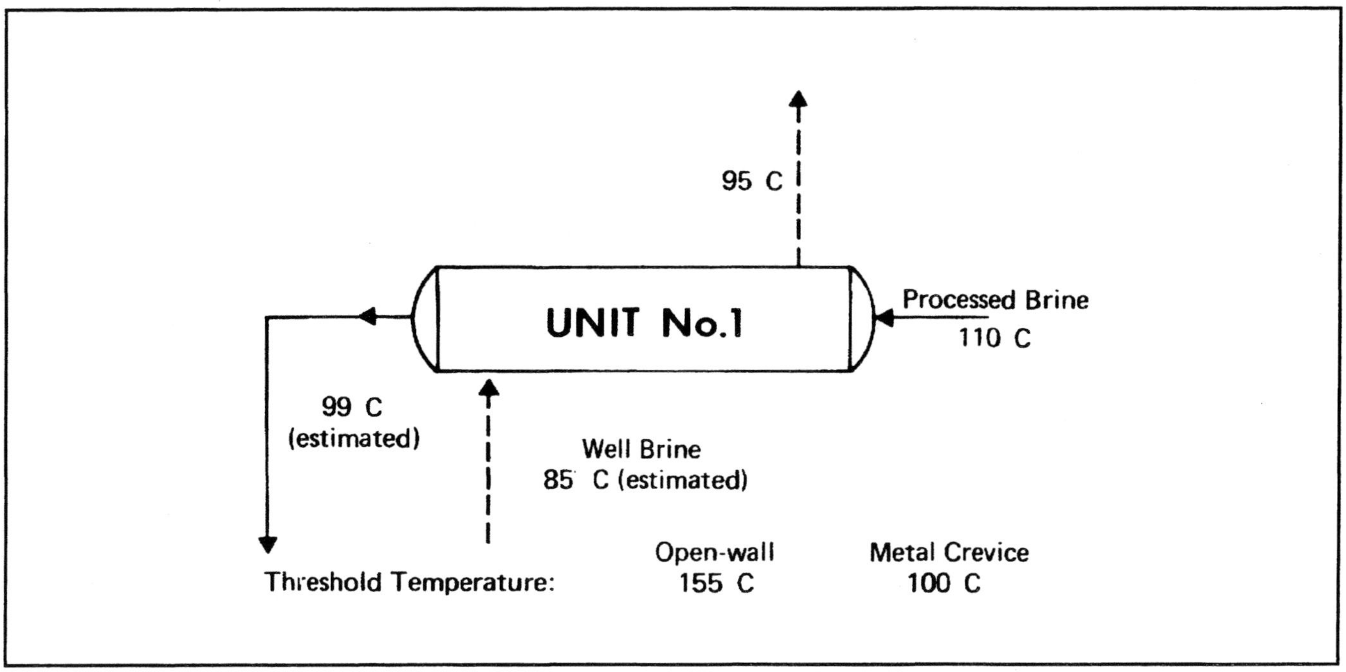

CASE HISTORY 2.30

Phenomenon: selective weld-metal attack
Group/ Type: nickel-based alloy UNS N10276 (alloy C-276)
Equipment: chlorine incinerator quench transition piece
Environment: sodium hypochlorate, NaCl, Cl_2, HCl
Conditions:
- (a) major species: 20% HCO_3 water
- (b) minor species: HCl
- (c) temperature: 250-500°F (121 to 260°C)
- (d) pressure: atmospheric
- (e) aeration: none

Time to failure: 30 days
Illustration: process schematic, photomacrograph, photomicrograph

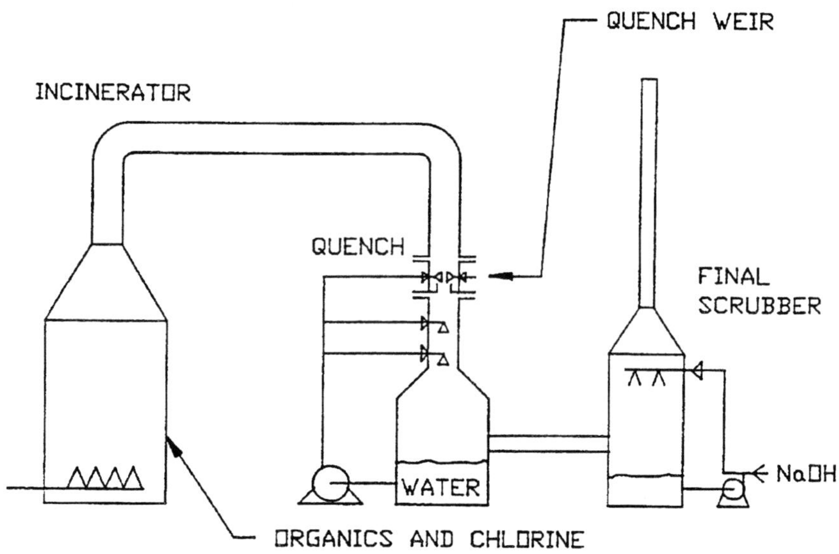

Comments: Quench water was used to cool hot gas coming off a brick-lined chlorine incinerator before the gas went into a brick-lined FRP column section (schematic). The quenching was done in a transition section made of SMA welded 1/8 inch thick alloy C-276. After 30 days service, the base material showed no serious evidence of attack. However, the welds were heavily corroded; measurement indicated a corrosion rate of 200 mpy (see photomacrograph). Most of the attack was concentrated along dendrite boundaries in the weld metal (see photomicrograph). X-ray microprobe analysis indicated the root pass of the welds had only 9.1% molybdenum, compared to the normal range of 15-17% molybdenum in the base material. It was concluded that the SMA weld flux had provided insufficient shielding to the point that relatively volatile molybdenum had become depleted in the weld deposit. The low-molybdenum content in the weld left it particularly prone to localized corrosion in this hot, very aggressive oxidizing environment.

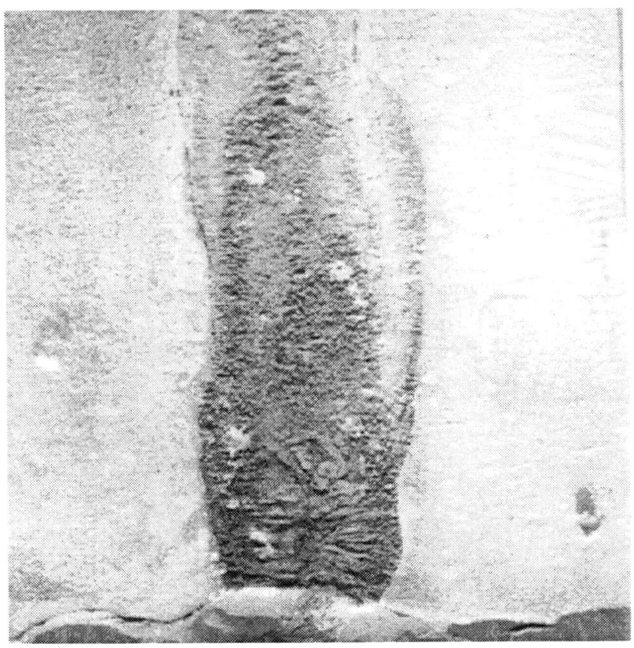

Figure 2.30(b) - Photomicrograph of weld deposit. Note corrosion along dendritic boundaries. 3X

Figure 2.30(c) - Photomacrograph of corroded weld deposit. 100X

CASE HISTORY 2.31

Phenomenon: catastrophic pitting under cathodic currents
Group/type: titanium
Equipment: chlor-alkali plant piping
Environment: chlorine-saturated NaCl solution
Conditions:
(a) major species: water, NaCl
(b) minor species: 260 g/L NaCl
(c) temperature: 194°F (90°C)
(d) pressure: not given
(e) aeration: not given
Time to failure: ~ 1 year
Illustration: macro
Comment: In a chlor-alkali plant using the electrolytic mercury-cathode process, titanium piping subjected to cathodic currents reportedly underwent intensive localized corrosion in welds and HAZs at rates of 6 mm/year (234 mils/year).
Remedy: An annular adapter, which created zones high in hypochorites and chlorates, was found to be effective in shielding the pipe terminals from cathodic current corrosion.

Reference: I.V. Riskin, V.B. Torshin, Ya. B. Skuratnik, M.A. Dembrovsky, *Materials Performance,* 22, 8 (1983): pp. 9-11.

CHAPTER 3

Galvanic Corrosion

CASE HISTORY 3.1

Phenomenon: galvanic corrosion, corrosion fatigue
Group/ Type: aluminum alloy, AA 443 casting (UNS A24430)
Equipment: cooling tower fan
Environment: cooling water mist
Conditions:

(a) major species: air
(b) minor species: water, 100% relative humidity
(c) temperature: ambient
(d) pressure: atmospheric
(e) aeration: saturated

Time to failure: in excess of 10 years
Illustration: fracture surface macro
Comment: During inspection of a cooling tower fan, an aluminum blade broke when struck by a worker's hammer. Examination showed that galvanic corrosion between the cast iron hub and the blade's shank had reduced the cross-sectional area of the shank from 6.3 in.^{2} to 1.4 in.^{2} A fatigue crack had initiated at the base of one of the corrosion pits (arrows, Figure). Discovery of the damage during inspection almost certainly prevented a catastrophic wreck of the fan during operation.
Remedy: Replace with fiberglass fan blades.

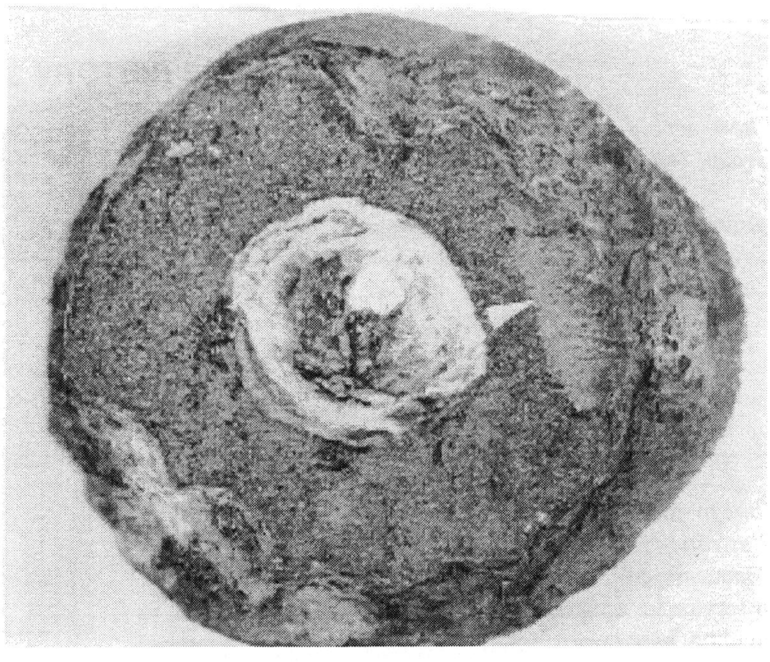

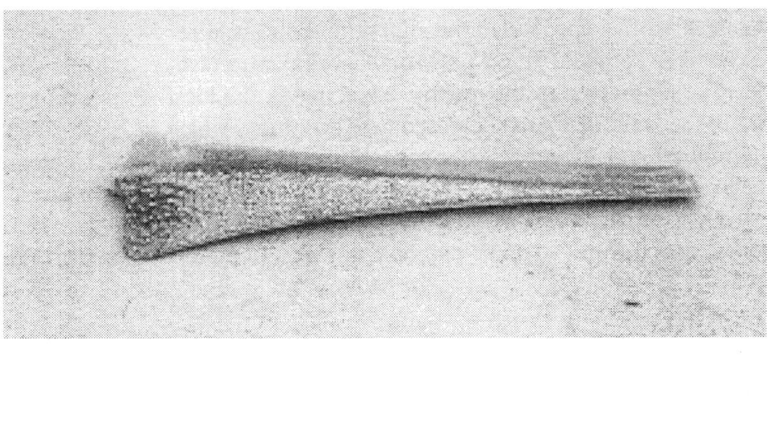

Figure 3.1 - (top) Fracture surface (bottom) as received.

Figure 3.2 - Macro

CASE HISTORY 3.2

Phenomenon: galvanic corrosion
Group/ Type: light metal aluminum alloy
Equipment: heat sensing actuator
Environment: industrial/marine atmosphere
Conditions:

(a) major species: water vapor, air
(b) minor species: chlorides
(c) temperature: Texas Gulf Coast ambient
(d) pressure: atmospheric
(e) aeration: saturated

Time to failure: 2.5 years
Illustration: macro
Comment: The aluminum barrel of this heat-sensing, fire-water actuator was coated with clear varnish to prevent corrosion. However, the design of the actuator allowed moisture to become trapped between the barrel and the bronze valve, resulting in rapid failure.
Remedy: Insulate dissimilar metal couples.

CASE HISTORY 3.3

Phenomenon: galvanic corrosion
Group/ Type: ferrous/AWS E6010 (UNS W06010) weld rod
quipment: slurry pipeline
Environment: brine slurry
Conditions:

(a) major species: 20% NaCl brine
(b) minor species: inert particles suspended in slurry
(c) temperature: 131°F (55°C)
(d) pressure: not given
(e) aeration: saturated

Time to failure: 3 years
Illustration: macro
Comment: AWS E6010 weld metal failed after three years in service in a schedule 40 (6 mm thick) ASTM A53 pipeline. Laboratory tests showed that the weld deposits were consistently ~ 30 mV anodic to the nearby base metal. With no galvanic effects, corrosion on the pipeline averaged ~ 30 mpy. When the slightly anodic weld deposits were coupled in an unfavorable area ratio (estimated as 64 to 1) with the nearby base metal, the local rate in the weld deposits increased to ~ 80 mpy, which was high enough to perforate the pipe at the welds in three years.
Remedy: Repair welds made with E7010-G weld electrodes, which were ~ 10 mV cathodic to the A53 pipe in this solution, have been in service for four years with no problems.

Figure 3.3 - Macro

Reference: E.L. Liening, "Trouble-shooting Industrial Corrosion Problems with Electrochemical Testing Techniques," Symposium on Non-destructive Testing and Electrochemical Methods for Monitoring Corrosion in Industrial Plants (Philadelphia, PA: ASTM, 1984).

CASE HISTORY 3.4

Phenomenon: galvanic corrosion
Group/ Type: ferrous/wrought iron
Equipment: structural support, statuary
Environment: marine atmosphere, New York harbor
Conditions:
(a) major species: air
(b) minor species: water vapor, chlorides, sulfur dioxide
(c) temperature: ambient
(d) pressure: atmospheric
(e) aeration: saturated

Time to failure: 100 years
Illustration: macro, schematic
Comment: The Statue of Liberty has a 3/32-in. thick outer skin of copper plate supported by a trusswork of 2 x 5/8-in. wrought iron bars. The outer skin is attached to the wrought iron bars with riveted copper saddles. The saddles were insulated from the bars with shellac-impregnated asbestos when the statue was erected. However, this insulation eventually failed, and over the years the wrought iron bars were heavily corroded by the combination of salt air, electrical contact with the more noble copper, and the collection of water beneath the saddles. The volume change when the iron turned to corrosion products warped the copper saddles and pulled many of the rivets out of the copper skin at 600 out of 1,500 saddles. With the statue's supporting structure dangerously weakened, major repair became mandatory.
Remedy: Replace wrought iron ribs with SS.
Reference: N. Nielsen, *Materials Performance*, 23, 4 (1984): pp. 78-82.

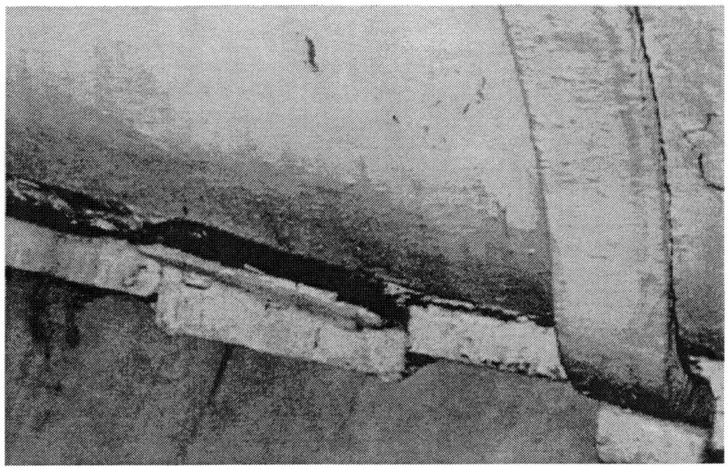

Figure 3.4 - Macro

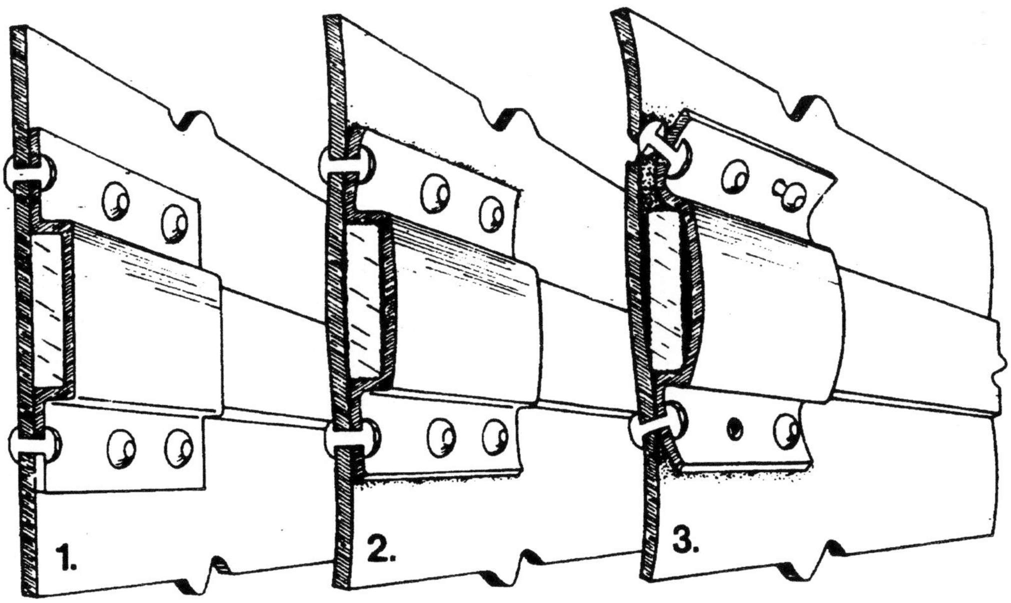

Figure 3.4 - Schematic

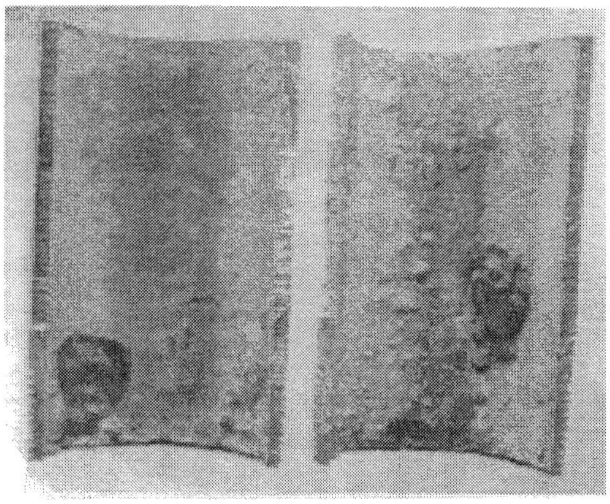

Figure 3.5 - Macro

CASE HISTORY 3.5

Phenomenon: galvanic corrosion
Group/ Type: ferrous, carbon steel
Equipment: piping
Environment: city water
Conditions:
(a) major species: water
(b) minor species: 50 ppm Cl-, carbonates, sulfates
(c) temperature: 140°F (60°C)
(d) pressure: 20 psig
(e) aeration: 6 ppm O_2
Time to failure: 18 months
Illustration: macro
Comment: Galvanized steel piping was used in a city water piping system in a commercial building. Pitting from oxygen cell corrosion was observed at many points throughout the system. However, at points where the galvanized steel piping was threaded into brass valves, attack was particularly severe. Galvanic coupling with the brass valves rapidly consumed the zinc coating. Once exposed, the steel pipe corroded at rates of ~ 60 mils/year in the areas in close contact with the brass.
Proposed Remedy: Insulate the brass valves, and inject a food-grade phosphate corrosion inhibitor into the water.

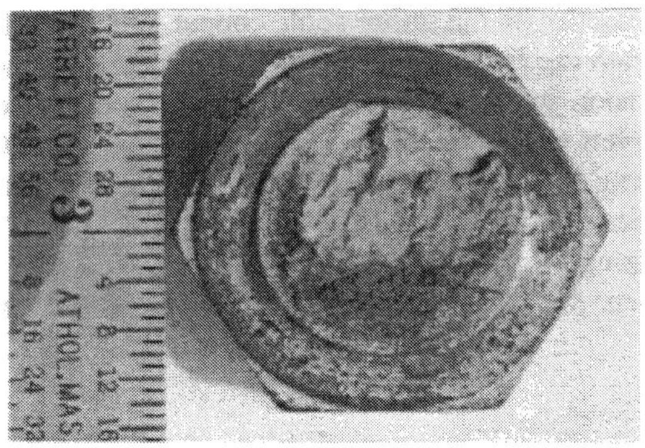

Figure 3.6(a) - Photomacrograph of failed bolt as received.

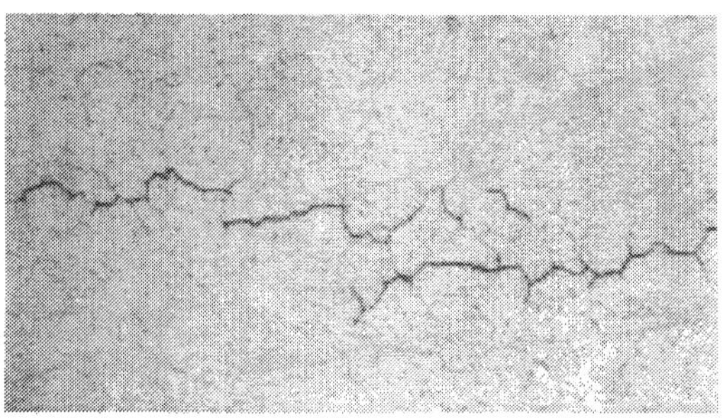

Figure 3.6(b) - Photomicrograph of cross section of failed bolt.

CASE HISTORY 3.6

Phenomenon: Galvanic coupling, leading to hydrogen embrittlement
Group/Type: Nickel-based alloy, Monel K-500 (UNS N05500)
Equipment: casing bolts, submersible pump, water injection well
Environment: aquifer water, pH = 5
Conditions:
a) major species: water, 137,000 mg/L chloride ion, 66,800 mg/L sodium ion
b) minor species: 9 mg/L H_2S, 306 mg/L CO_2
c) temperature: 130ºF (54°C)
d) pressure: 3,000 psig
e) aeration: < 5 ppb O_2
Time to failure: 37 days
Illustration: macro, micro
Comment: Monel K-500 (UNS N05500) bolts were used to assemble a CDA 958 Ni-Al bronze pump. The pump failed after 37 days in service. On disassembly it was found that one of the bolts had fractured, showing a progressive fracture due to environmental cracking. The fracture showed considerable crack branching and had an intergranular mode. Hydrogen embrittlement due to galvanic coupling with the bronze pump casing was considered the most likely cause of failure.
Remedy: Replace the bolts with duplex stainless steel or precipitation-hardening Cupro-nickel bolts.

CASE HISTORY 3.7

Phenomenon: galvanic corrosion
Group/ Type: unalloyed molybdenum
Equipment: positive terminal on batteries
Environment: ambient temperature air
Conditions:

(a) major species: air
(b) minor species: water vapor
(c) temperature: ambient
(d) pressure: atmospheric
(e) aeration: saturated

Time to failure: 3 months
Illustration: macro
Comment: Although molybdenum is immune to atmospheric corrosion in all manner of environments, severe general wastage of molybdenum positive-terminal pins occurred on a number of three-volt lightweight lithium batteries after three to six months in ambient temperature storage. Atmospheric moisture had condensed on the battery caps. If enough condensate formed, a conductive current path occurred between the battery's positive pin and negative case. When polarized to an anodic potential of -3.0 volts, even molybdenum pins were rapidly consumed by this formidable galvanic corrosion cell.
Remedy: Cover the terminals with insulating, moisture-impenetrable plastic outer casings for field storage.
Contributor: Dr. W.R. Cieslak, Corrosion Division, Sandia National Laboratories

CHAPTER 4

Environmental Cracking

CASE HISTORY 4.1

Phenomenon: SCC
Group/ Type: light metals, aluminum alloy; AA 7075-T6 (UNS A97075)
Equipment: aircraft hydraulic cylinder
Environment: ambient shop air
Conditions:
- (a) major species: ambient air
- (b) minor species: water vapor, 60 to 80% relative humidity
- (c) temperature: ambient
- (d) pressure: atmospheric
- (e) aeration: total

Time to failure: 3 weeks
Illustration: macro of fracture surface, SEM fractograph
Comments: This hydraulic cylinder was one of a lot mistakenly heat-treated to the high strength T6 condition rather than the overaged, lower-strength T73 condition. The parting plane of high-strength aluminum alloy forgings is particularly sensitive to SCC even in rather innocuous environments. This particular hydraulic cylinder cracked after installation, but before any flight time; therefore, its entire "service life" consisted of sitting in part bins and on assembly lines in a midwestern aircraft assembly plant. Interference fitting a steel attachment pin to the hydraulic cylinder resulted in residual tensile stresses around the pin hole. The cylinder had been anodized and painted after heat treatment to protect it against corrosion. However, when the pin was shrunk into its hole, the tensile stresses imposed on the surrounding aluminum cracked the anodized coating, leaving defects where corrosion could occur. Galvanic coupling with the steel pin probably accelerated the attack.
Remedy: This heat treat lot of cylinders was re-heat-treated to the T73 condition. No further failures occurred.

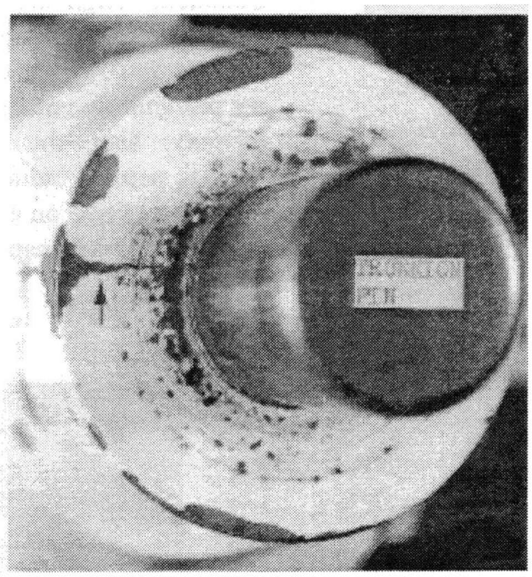

Figure 4.1(a) - (above) Crack in cylinder (arrow). 3X

Figure 4.1(b) Exposed fracture surface. 3X

Figure 4.1(c) - Photomicrograph at origin, 100X.

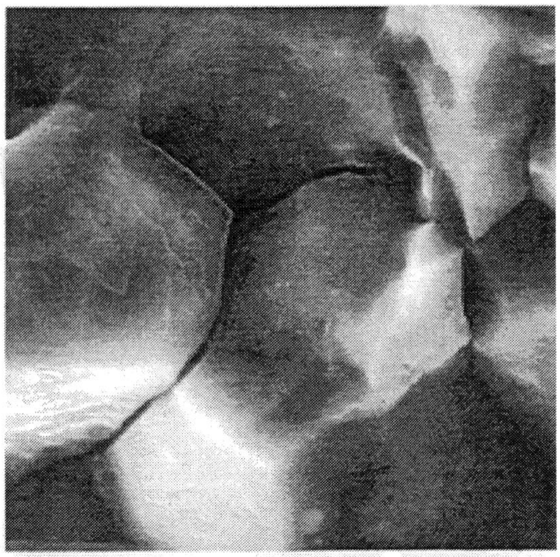

Figure 4.2 - SEM fractograph

CASE HISTORY 4.2

Phenomenon: SCC
Group/ Type: light metals; aluminum alloy, AA 7079-T652 (UNS A97079)
Equipment: aircraft main wing spar
Environment: marine atmosphere
Conditions:
(a) major species: air
(b) minor species: water vapor, NaCl
(c) temperature: tropical ambient
(d) pressure: 1 atm
(e) aeration: natural
Time to failure: not given
Illustration: SEM fractograph
Comment: High-strength aluminum alloys in the fully hardened condition are susceptible to SCC in the salt air near seacoasts. Aluminum structurals for naval aircraft are particularly vulnerable to this type of attack.
Remedy: Use the overaged T7 condition when strength constraints permit. Induce residual compressive stresses (shot-peen, interference fits) on exposed surfaces. Anodize, seal, and coat exposed surfaces. Inspect critical items frequently.

CASE HISTORY 4.3

Phenomenon: IGSCC
Group/ Type: light metals; AA 5083-0 with AA 5183 (UNS A95183) weld metal
Equipment: 24 in. diameter pipe flange
Environment: atmospheric
Condition:
(a) major species: ambient air
(b) minor species: water vapor, chlorides
(c) temperature: ambient
(d) pressure: atmospheric
(e) aeration: saturated
Time to failure: 3.5 years total
Illustration: cross-sectional views, micro and macro
Comments: This pipe flange was heated to above 300°F (149°C) for a total of ~ 50 hours during the course of shop welding and field weld repairs. The flange was stored at a seacoast location for seven months before being welded into its pipe. Seven months after welding, the line was insulated, and eight months after insulation, the plant was started up. After 18 months at cryogenic temperatures, cracks were discovered in both the weld metal and HAZs. As the photomicrographs show, the preheat had sensitized the flange, causing grain boundary precipitates to form, which allowed IGSCC when the pipe was exposed to the chloride-bearing seacoast atmosphere. The attack was assumed to occur during the period when the insulation caught and held atmospheric moisture against the pipe surface before start-up. It was assumed that no attack would occur at the cryogenic temperatures of normal service.
Remedy: Avoid preheat times in excess of 6-8 h at 93 to 232°C.
Contributor: D.N. Williams, Battelle Columbus Laboratories.

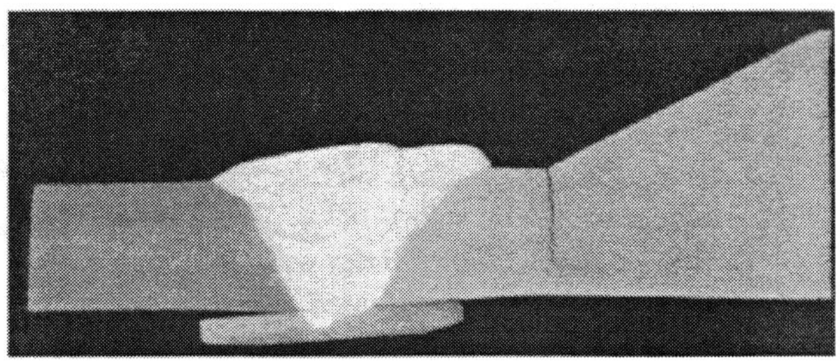

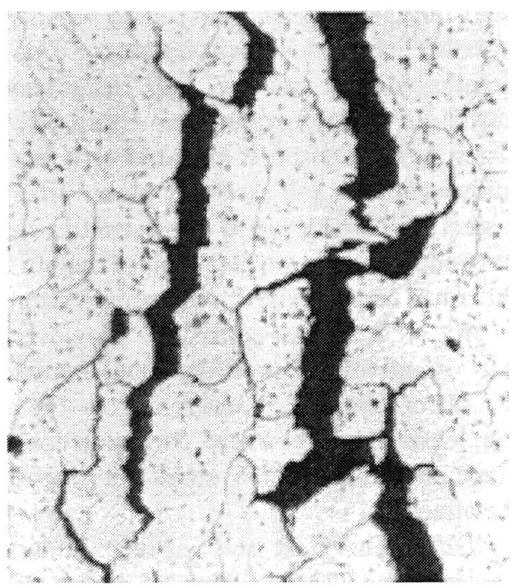

CASE HISTORY 4.4

Phenomenon: corrosion-induced hydrogen embrittlement
Group/Type: ferrous, AISI 4140 (UNS G41400) steel, HRC 36, cadmium-plated
Equipment: bonnet bolt on 600-lb steam valve
Environment: atmospheric, marine-industrial
Conditions:
(a) major species: air
(b) minor species: water vapor, chlorides
(c) temperature: 500°F (260°C) steady state, ambient during shut-downs
(d) pressure: 1 atm
(e) aeration: atmospheric
Time to failure: 10 years
Illustration: fracture surface, photomicrograph
Comment: The cadmium-plating widely used for corrosion protection on low strength carbon steel bolts for mild atmospheres has serious limitations for more severe marine and marine-industrial uses. The cadmium is consumed more quickly than zinc in humid, chloride-bearing atmospheres, and hydrogen generated by the corrosion can embrittle high strength steel substrates. This bonnet bolt fractured by hydrogen embrittlement after 10 years in service, releasing its stored strain energy with enough violence to injure a nearby worker. Examination showed the slightly branched intergranular fracture mode characteristic of hydrogen embrittlement. At the point of origin, the cadmium plating had corroded away, exposing cathodic base metal, which would absorb hydrogen. Since hydrogen embrittlement is rare at temperatures above ~ 150°F (66°C), it was postulated that no hydrogen cracking occurs at the relatively high steady-state service temperature. The long service time suggests that cracking occurs only at near-ambient temperatures during occasional unit shutdowns.
Preventive measures: Replace with TFE-fluorocarbon-coated steel or precipitation-hardening SS bolts.

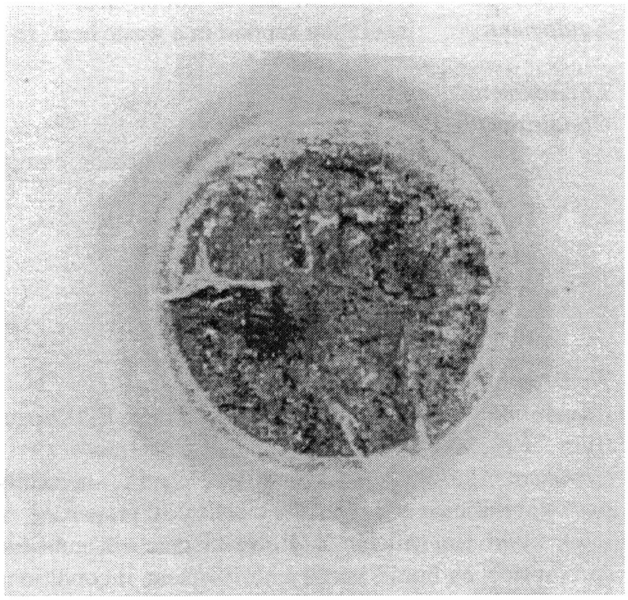

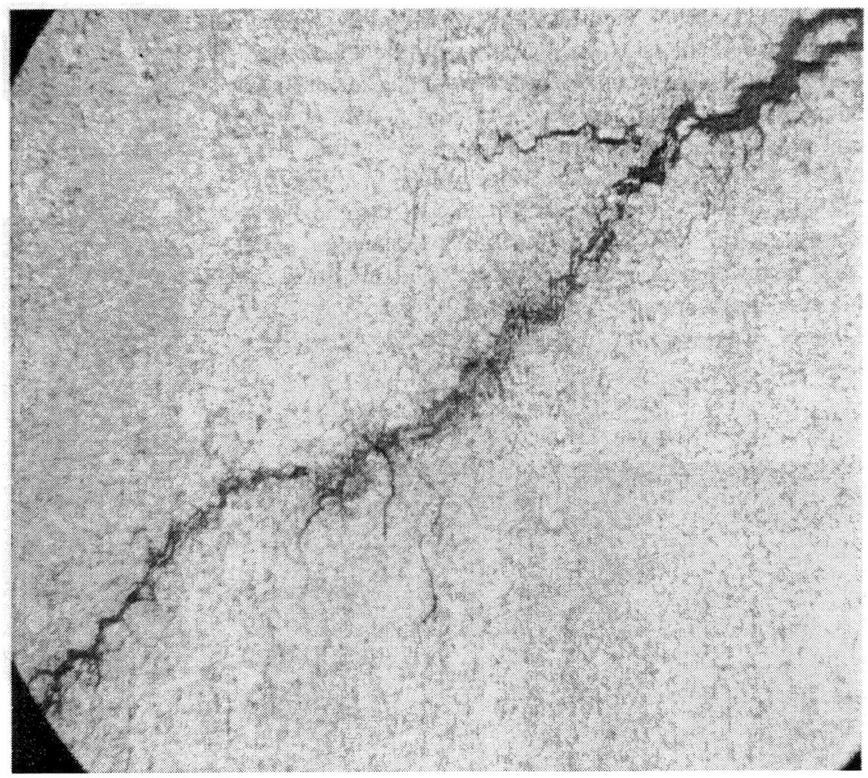

Figure 4.4 - (a) Photomicrograph and (b) fracture surface.

CASE HISTORY 4.5

Phenomenon: liquid metal embrittlement
Group/ Type: ferrous/galvanized carbon steel
Equipment: insulation support in a waste heat recovery unit
Environment: turbine exhaust gas
Conditions:
(a) major species: carbon monoxide, carbon dioxide
(b) minor species:
(c) temperature: 700 to 800°F (371 to 426°C)
(d) pressure: atmospheric
(e) aeration: anaerobic
Time to failure: 10 years
Illustration: macro, micro, SEM fractograph, EDX spectrum
Comment: Hot-dip galvanizing (zinc-coating) normally has no significant effect on the mechanical properties of low-carbon steel structurals. However, zinc will embrittle carbon steel by liquid metal embrittlement if conditions are right. After 10 years in service at 700 to 800°F (371 to 426°C), the galvanized steel insulation support was so brittle pieces could be broken off by hand. SEM examination showed that the fracture was entirely intergranular. A thin grain-boundary film was present throughout the microstructure on metallographic examination. EDX examination revealed high levels of zinc on the fracture surface. The melting point of zinc is 784° F (417.7°C); exposure to temperatures in excess of that will allow the molten zinc to diffuse into the steel substrate with time. Zinc in the grain boundaries will cause embrittlement, even on low stress items, such as this insulation support.
Remedy: Replace with uncoated AISI 304 SS (UNS S30400).

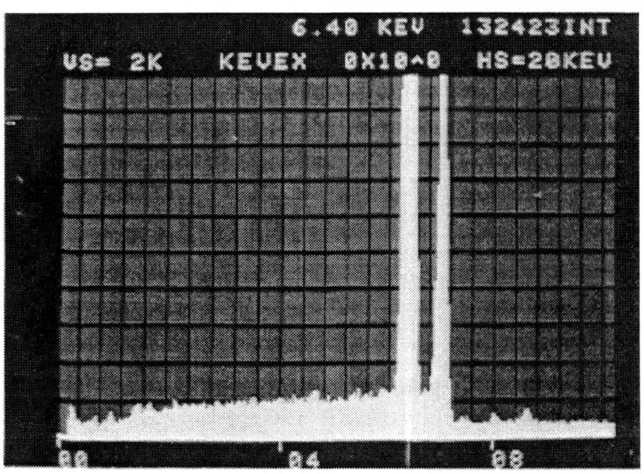

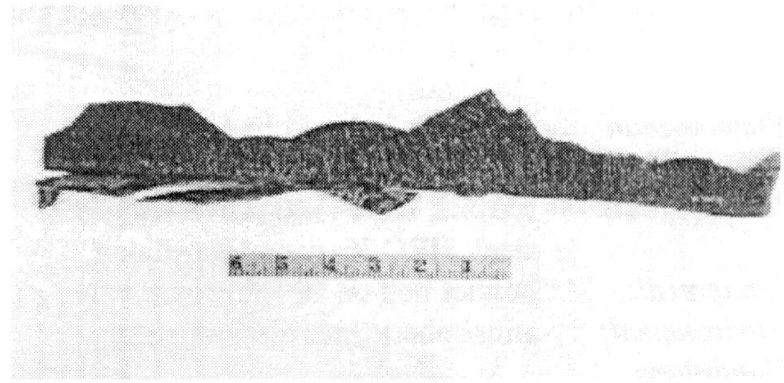

Figure 4.5(a)

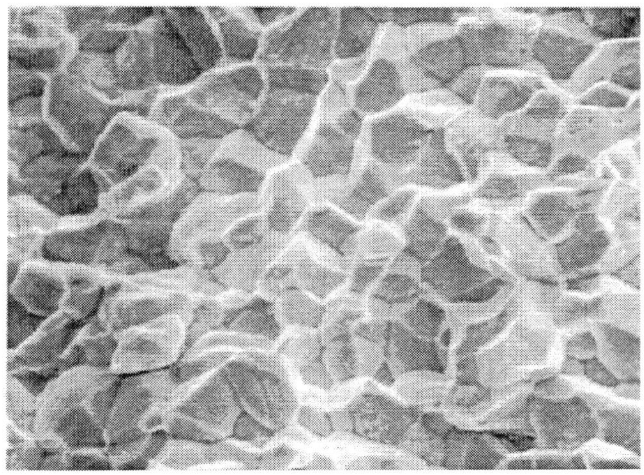

Figure 4.5(b)

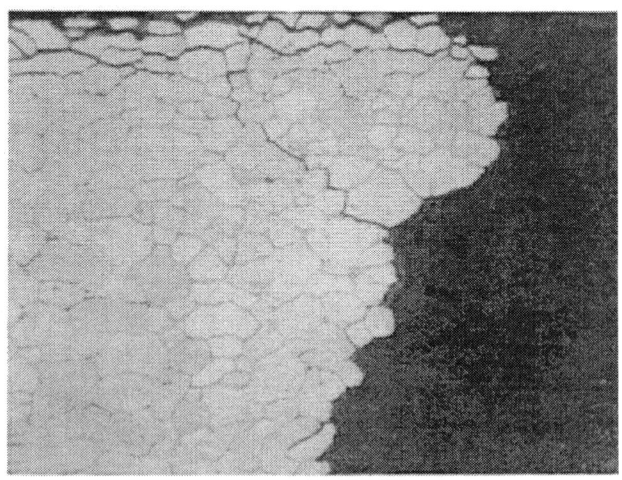

Figure 4.5(c)

Figure 4.5 - Macro (a); micro (b); SEM fractograph (c); and (d) EDX spectrum.

CASE HISTORY 4.6

Phenomenon: liquid metal attack
Group/ Type: ferrous, AISI 4330V alloy steel
Equipment: aircraft hydraulic cylinder
Environment: ambient air
Conditions:
(a) major species: tin plate
(b) minor species: none
(c) temperature: atmospheric ambient
(d) pressure: atmospheric
(e) aeration: complete
Time to failure: less than 3 months
Illustration: photomicrograph, SEM micrograph
Comments: The attachment lugs of this high strength steel hydraulic cylinder were tin-plated in an attempt to cut down on fretting between the cylinder and the adjacent structure. However, the frictional heat from vibration in service was sufficient to melt the tin and produce intergranular penetration along prior austenitic grain boundaries. Such intergranular penetration is highly undesirable since it can produce early failure by fatigue in the highly stressed cylinder.
Remedy: Use a baked-on molybdenum disulfide coating instead of tin-plating.

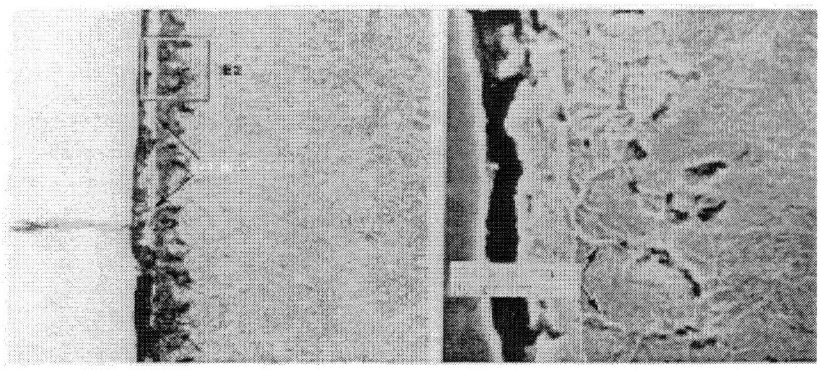

Figure 4.6 - Photomicrograph and SEM micrograph, 400X and 1,800 X.

CASE HISTORY 4.7

Phenomenon: Hydrogen stress cracking
Group/ Type: ferrous; AISI 4037 (UNS G40370), hardness 448 KHN-500g
Equipment: fastener
Environment: mild industrial atmosphere
Conditions:
(a) major species: air
(b) minor species: water vapor
(c) temperature: ambient
(d) pressure: atmospheric
(e) aeration: 20% by volume
Time to failure: 3 weeks
Illustration: macro, SEM fractograph
Comment: This cadmium-plated steel fastener failed in an assembly line before its structure was completed. SEM examination showed intergranular facets overlaid with light etching from corrosion. The hardness exceeded specifications, and the alloy was found to be a carbon-1/4 molybedum grade rather than the specified chromium-nickel-molybdenum 4340 grade.
Remedy: Verify hardness and alloy before assembly. Induce compressive residual stresses when possible using shot-peening, rolled threads, etc.

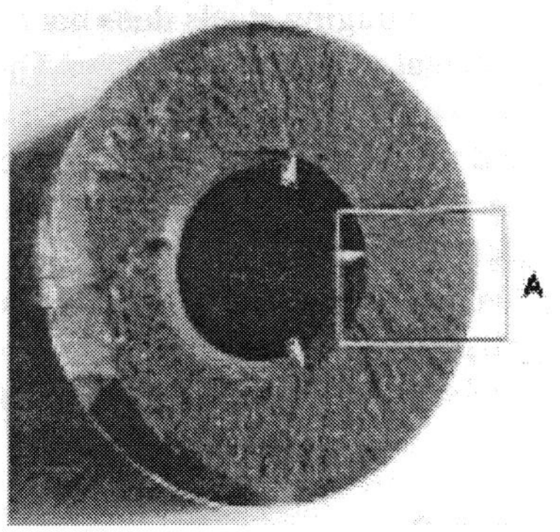

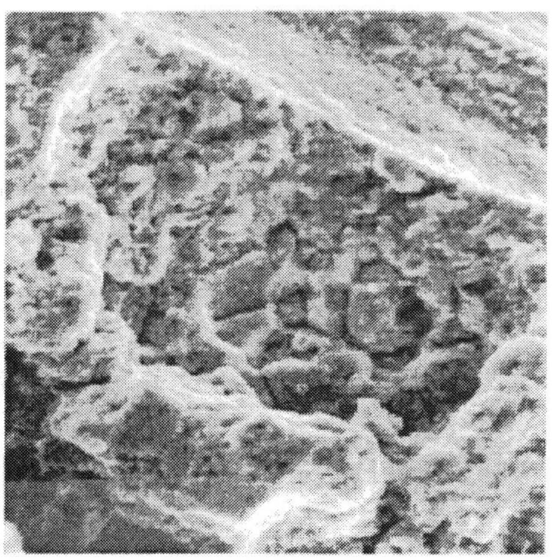

Figure 4.7- Macro and SEM fractograph.

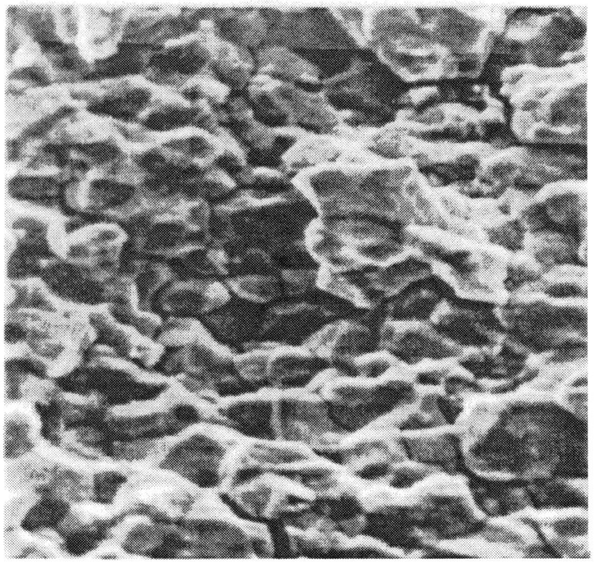

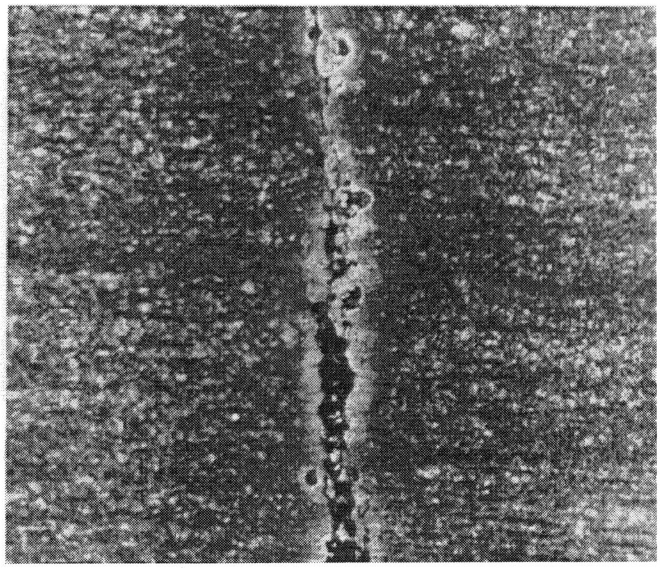

Figure 4.8 - (a) SEM fractograph (2,000X) and (b) photomicrograph (250X).

CASE HISTORY 4.8

Phenomenon: hydrogen-induced SCC
Group/ Type: ferrous, maraging steel, 250 ksi YS
Equipment: rocket engine gas generator
Environment: mild industrial atmosphere
Conditions:
(a) major species: air
(b) minor species: water vapor, SO_2
(c) temperature: ambient
(d) pressure: atmospheric
(e) aeration: natural
Time to failure: not given
Illustration: SEM fractograph, optical photomicrograph
Comment: The exceptionally high strength and toughness of maraging steels does not exempt them from environmental cracking problems. This gas generator vessel passed a hydrostatic test after fabrication, then burst when filled for service. Moisture left from the hydrotest and in shop air initiated corrosion, which generated enough hydrogen to crack this material.
Remedy: Induce residual compressive stresses (shot-peen or autofrettage). Use steel with YS below 180 ksi when strength requirements permit.

CASE HISTORY 4.9

Phenomenon: corrosion fatigue
Group/ Type: ferrous/AISI 4340 (UNS G43400) alloy steel, HRC 35
Equipment: gas intensifier cylinder
Environment: industrial atmosphere
Conditions:
(a) major species: air
(b) minor species: water vapor (relative humidity averages 60 to 70%)
(c) temperature: -140 to 150°F (-96 to 66°C)
(d) pressure: atmospheric
(e) aeration: saturated
Time to failure: 1 year
Illustration: macros (Figures 4.9a and 4.9b, page 51)
Comment: During normal operation, this intensifier cylinder operates with a high-pressure hydrocarbon gas on the bore. However, during shutdowns, water vapor condensed in the cylinders, allowing random pitting. Fatigue cracks initiated form these pits during normal pressure cycling.
Remedy: Inject a volatile inhibitor into the cylinders during shutdowns.

CASE HISTORY 4.10

Phenomenon: internal hydrogen embrittlement
Group/ Type: ferrous, AISI alloy 4340 (UNS G43400), HRC 36 to 40
Equipment: oil well drilling tool
Environment: fabrication processing
Conditions:

(a) major species: zinc phosphate, phosphoric acid
(b) minor species: stannate inhibitors
(c) temperature: 175°F (79°C)
(d) pressure: atmospheric
(e) aeration: not controlled

Time to failure: 1 week to 2 months
Illustration: macro, SEM fractograph
Comment: Tons of oil country tubular goods are phosphatized each year. If the hardness is low (below about HRC 32) and the bath chemistry is in balance, no problems normally result. However, this material had a hardness of HRC 36, and the zinc phosphating bath was not closely monitored for adequate inhibitor and iron content. One batch of parts failed in service, although this particular item is never exposed to well-bore fluids. Analysis showed the classic pattern of intergranular cracking with subsurface initiation, which is characteristic of internal hydrogen embrittlement.
Remedy: Monitor inhibitor and iron content in phosphatizing baths and test treated parts for delayed failure.

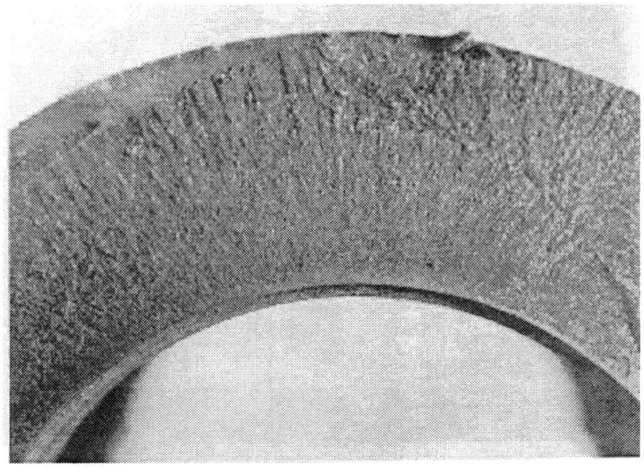

Figure 4.10(a) - Macro

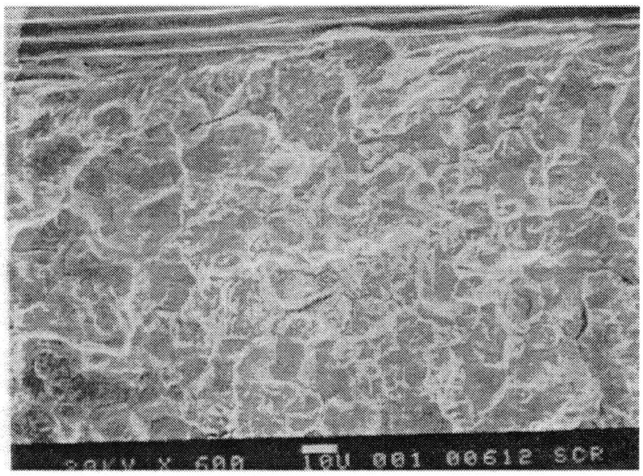

Figure 4.10(b) SEM fractograph

Figure 4.9(a) - Inner cylinder surface 10X.

Figure 4.9 (b) - Cross section at inner surface. Fracture surface is at top (100X).

Figure 4.11(a) - Collapse of the Point Pleasant Bridge.

Figure 4.11(b) - Macro

CASE HISTORY 4.11

Phenomenon: SCC and corrosion fatigue
Group/Type: ferrous/ ASTM A7 structural steel
Equipment: bridge suspension eyebar
Environment: semi-industrial atmosphere
Conditions:
(a) major species: air, water vapor
(b) minor species: hydrogen sulfide
(c) temperature: ambient
(d) pressure: 1 atm
(e) aeration: saturated
Time to failure: 40 years
Illustration: macro
Comment: Forty-six people died in the collapse of the Point Pleasant Bridge in West Virginia. The National Transportation Safety Board concluded that a hydrogen sulfide stress corrosion crack initiated the brittle fracture of a critical eye-bar. Contributing factors included high steel hardness, a designed-in crevice where water could collect, and local stresses greater than the yield strength of the steel.
Remedy: Control hardness, protect creviced areas with coatings applied before assembly, and lower design loads in critical areas.
Reference: "Collapse of U.S. 35 Highway Bridge, Point Pleasant, West Virginia," NTSB Report No. NTSB-HAR-71-1, December 1967.

CASE HISTORY 4.12

Phenomenon: liquid metal attack
Group/Type: ferrous, ASTM A516 grade 70 (UNS K02700) carbon steel
Equipment: pressure vessel
Environment: fabrication
Conditions:
(a) major species: liquid 70Cu-30Ni alloy
(b) minor species:
(c) temperature: 2,200+ °F (1,216+ °C)
(d) pressure: atmospheric
(e) aeration: MIG shield
Time to failure: immediate
Illustration: micro
Comment: The copper-nickel alloys have been welded directly to carbon steels in many noncritical applications. However, grain boundary penetration of copper into the steel (Figure) is a possibility that should be considered when designing highly stressed welds.

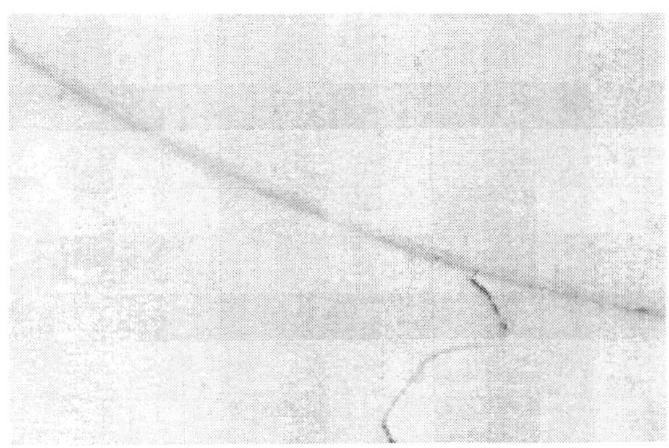

Remedy: Adjust welding technique to minimize heat input, thus minimizing the depth of grain boundary penetration.

CASE HISTORY 4.13

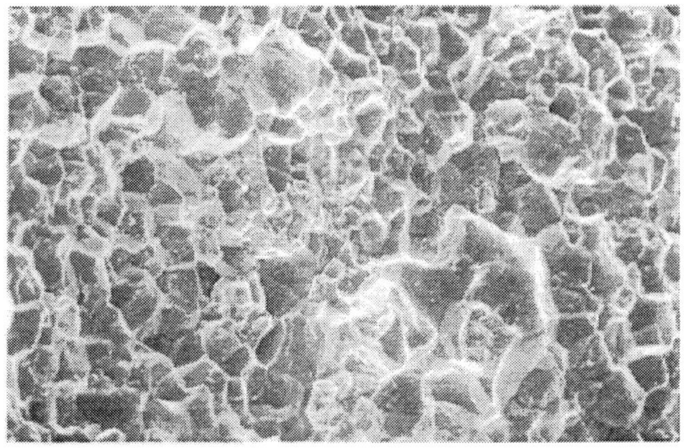

Phenomenon: liquid metal cracking
Group/ Type: ferrous, low alloy steel bolting ASTM A193, grade B7
Equipment: valve bonnet bolt
Environment: industrial atmosphere
Conditions:

(a) major species: air, cadmium
(b) minor species: none
(c) temperature: 730°F (387°C)
(d) pressure: atmospheric
(e) aeration: restricted

Time to failure: less than 1 year
Illustration: SEM fractograph
Comment: The valve bonnet on a steam valve blew off when all four bonnet bolts failed catastrophically. These bolts had been cadmium-plated for atmospheric corrosion resistance before installation and during shutdowns. However, the operating temperature was significantly above the melting point of cadmium [609.8°F (321°C)]. SEM examination revealed a completely intergranular fracture path, and EDX analysis showed a cadmium concentration of 26% on the intergranular facets.
Remedy: Use bolts with a black oxide conversion coating rather than a sacrificial metallic coating.

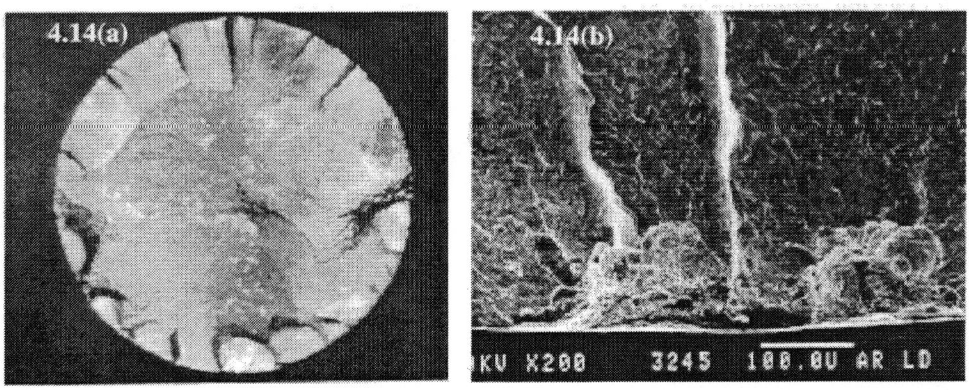

Figure 4.14-
(a) fracture surface
(b) SEM fractograph
(c) photomicrograph

CASE HISTORY 4.14

Phenomenon: corrosion fatigue and SCC
Group/ Type: ferrous, low alloy steel, AISI 4340, HRC 44, chromium-plated
Equipment: aircraft landing gear pins
Environment: Arabian Gulf seacoast atmosphere
Conditions:

a) major species: air
b) minor species, sea salt, sand, water vapor
c) temperature: Arabian Gulf ambient (32 to 120°F [0 to 49°C])
d) pressure: 1 atmosphere
e) aeration: natural

Time to failure: 8 years (5,000 flight hours)
Illustration: Fracture surface macro, SEM fractograph, photomicrograph
Comment: A cargo aircraft lost steering control on landing. Examination revealed that the four steel pins in the nose wheel gear had fractured. Beachmarks and fatigue

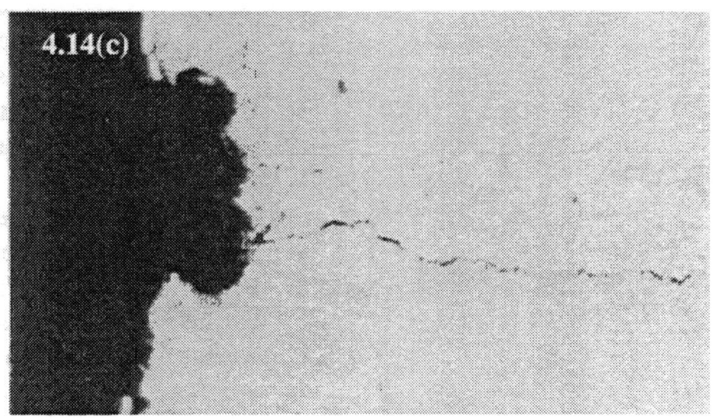

striations indicated a fatigue failure initiating at corrosion pits, with some intergranular facets due to SCC by corrosion-generated hydrogen.
Remedy: Decrease inspection interval to 1,200 flight hours (every 2 years), and use electroless nickel plating for increased corrosion resistance compared to chromium plating.

CASE HISTORY 4.15

Phenomenon: SCC
Group/ Type: ferrous, ASTM A285 grade C (UNS K02801) carbon steel, C 0.067%
Equipment: boiler feedwater deaerator
Environment: steam, deionized water
Conditions:
(a) major species: deionized water
(b) minor species: calcium
(c) temperature: 240°F (115°C)
(d) pressure: 25 psig
(e) aeration: deaerated
Time to failure: 5 years
Illustration: macro, micro
Comment: Unlike most cracking in boiler feedwater deaerators, (usually attributed to corrosion fatigue) this crack in a steam inlet nozzle showed branched, intergranular cracking, indicative of SCC.
Remedy: Thermal stress relief for new construction.

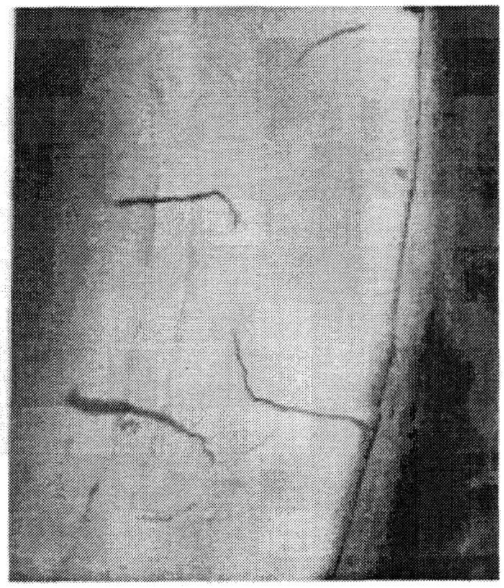

Figure 4.15 - (a) Original cracking inside vessel and (b) cross-section, 400X.

CASE HISTORY 4.16

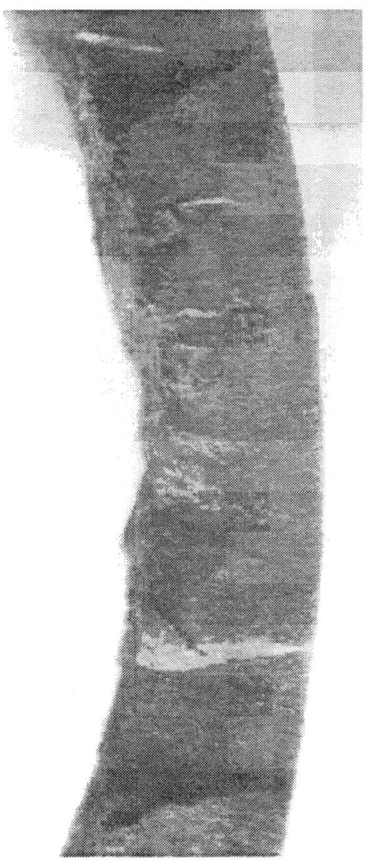

Phenomenon: corrosion fatigue
Group/ Type: ferrous, AISI 4130 (UNS G41300)
Equipment: drill pipe
Environment: seawater
Conditions:
(a) major species: water, 3.5% NaCl
(b) minor species: oxygen, carbonates
(c) temperature: ambient
(d) pressure: hydrostatic at depths up to 5,000 meters
(e) aeration: saturated
Time to failure: 7 years
Illustration: macro
Comments: After seven years of service, a string of 4.5-in. diameter by 0.375-in. wall S-135 drill pipe parted during a deep-sea drilling operation. This string was coated with inorganic zinc paint, but only on the outside. Seawater was circulated on the inside to lubricate the drilling and remove the cuttings. The string had been inspected within two months of failure with a magnetic flux-leakage unit; no defects were detected in the joint that failed. Examination of the fractured joint revealed that extensive general corrosion on the inside surface of the pipe wall had consumed as much as 40% of the nominal wall thickness. No beachmarks showed on the fracture surface, but the fractures were perpendicular to the longitudinal axis of the pipe, transgranular, and non-branching. Radial marks left by shear steps between crack fronts indicated crack initiation on the inside surface, at the bottom of corrosion pits. Therefore, the failure was attributed to corrosion fatigue initiating in areas of high stress caused by thinning of the wall by general corrosion.
Remedy: Use pipe coated on both inside and out, and improve inspection so corroded pipe will be retired before more than 12% of the wall thickness is consumed.

CASE HISTORY 4.17

Phenomenon: corrosion fatigue
Group/ Type: ferrous, API 5LX-52 carbon steel line pipe
Equipment: marine riser
Environment: seawater
Conditions:
(a) major species: H_2O, 3%NaCl (Gulf of Mexico seawater)
(b) minor species: O_2, $CaCO_3$, etc.
(c) temperature: 70 to 82°F (21.1 to 27.8°C)
(d) pressure: water depth - 8 ft
(e) aeration: saturated
Time to failure: 25 days
Illustration: SEM fractograph
Comment: This SEM fractograph from a reversed bending fatigue failure in seawater illustrates how the fatigue striations normally produced by high-cycle fatigue are coarsened and overlaid with corrosion products when corrosion fatigue operates.
Remedy: Reduce mean stress; apply cathodic protection.

Figure 4.17 - SEM fractograph

CASE HISTORY 4.18

Phenomenon: corrosion fatigue
Group/Type: ferrous, carbon steel, STM A285 grade C (UNS K02801)
Equipment: boiler feedwater deaerator
Environment: softened boiler feedwater
Conditions:
(a) major species: water
(b) minor species: CO_2, iron scale
(c) temperature: 220°F (104°C)
(d) pressure: 15 psig
(e) aeration: saturated on inlet, de signed for 7 ppb outlet (outlet O_2 not analyzed)
Time to failure: 30 years
Illustration: macro
Comment: Weld or HAZ cracking is encountered in 20 to 40% of welded steel boiler feedwater deaerators. Causes are currently not well understood, but cracking is apparently related to welding residual stresses.
Remedy: Repair weld and use thermal stress relief.

Figure 4.18 - Macro of cracks in deaerator storage tank weld. The dark areas are surfaces of the corrosion fatigue cracks on either side of weld, revealed by breaking the sample open.

Source: D.C. Bennett, "Corrosion Problems in the Pulp and Paper Industry," *Materials Performance*, 1(1984): pp. 9-15.

CASE HISTORY 4.19

Phenomenon: corrosion fatigue
Group/ Type: ferrous, AISI 4340 (UNS G43400), HRC 35
Equipment: process tubing
Environment: steam condensate
Conditions:.
(a) major species: water
(b) minor species: CO_2 (up to 40 ppm)
(c) temperature: 480°F (249°C)
(d) pressure: 600 psi
(e) aeration: deaerated
Time to failure: 7 years
Illustration: as-received and fracture surface macros
Comment: Internal pressure pulsations produced a maximum stress of 7.5 ksi on the outer surface of these tubes, at a rate of 30 cpm. Such a low cyclic stress is not normally considered harmful. However, the cold-bending in fabrication imposed a residual tensile stress calculated as 126 ksi (87% of the yield stress). Hot water high in CO_2 yet low in oxygen can produce both SCC and corrosion-fatigue crack propagation rates higher than either CO_2-free water or highly oxygenated water. The result, in this case, was unexpected failure by corrosion fatigue at high mean stress and low alternating stresses.
Remedy: Eliminate residual fabrication stresses by thermal stress relief.

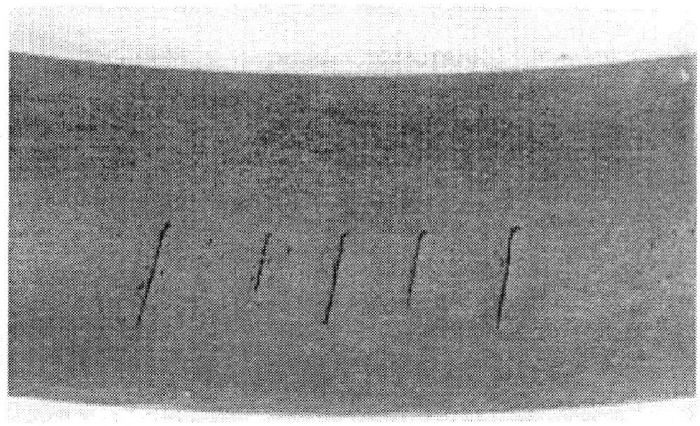

Figure 4.19 - (a)(top) as received and (b)(above) fracture surface macros.

CASE HISTORY 4.20

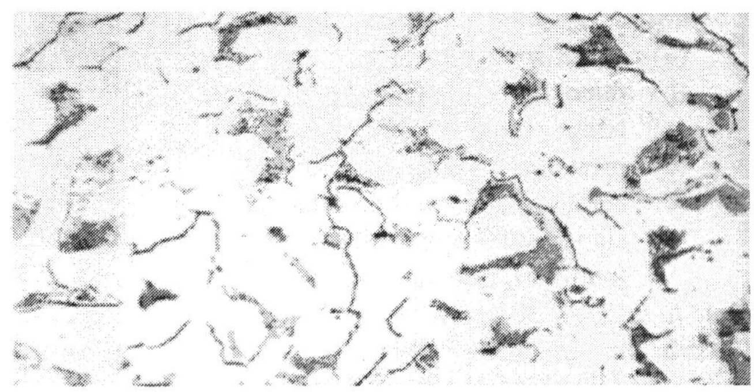

Phenomenon: Hydrogen attack
Group/ Type: ferrous, carbon steel
Equipment: nose section and roof tubes, utility boiler
Environment: steam
Conditions:
(a) major species: water, steam
(b) minor species: traces of NaOH, coordinated phosphate treatment
(c) temperature: not given
(d) pressure: 2,075 psig
(e) aeration: deaerated
Time to failure: 25 years
Illustration: macro, micro (right and far right)
Comment: Caustic gouging typically produces a line of local corrosion. However, in boilers operating above 1,000 psig, hydrogen can be produced by the reaction, $Fe + 2\ NaOH = Na_2FeO_2 + H_2$. In this case, hydrogen evolved during caustic gouging caused severe fissuring and rupture of roof tubes in a high pressure, forced circulation boiler in peaking service.
Remedy: not given
Reference: R.D. Port, Nalco Chemical Co.

CASE HISTORY 4.21

Phenomenon: corrosion fatigue
Group/ Type: ferrous, API 5L gr. B carbon steel
Equipment: 3 in. diameter boiler blowdown piping
Environment: boiler water
Conditions:

a) major species: water, conductivity 460-1,385 micromhos, pH 10-11.5
b) minor species: 2-4 ppm silica, total dissolved solids 300-900 ppm
c) temperature: cyclic, ambient to 500ºF (260°C)
d) pressure: cyclic, 1 atm to 600 psig
e) aeration: unknown

Time to failure: 8 years
Illustration: photograph, photomicrograph
Comment: Every day, a boiler dumped 5,000 gallons of water to an atmospheric blowdown drum through a run of 3 in. diameter API 5L gr. B piping. After eight years service, a pipe elbow began leaking. Examination indicated a multitude of cracks on the inner surface (photograph). These cracks were transgranular, non-branching, blunt-tipped and oxide-filled, indicative of corrosion fatigue.
Remedy: Increase the pipe schedule to decrease pressurization stresses.

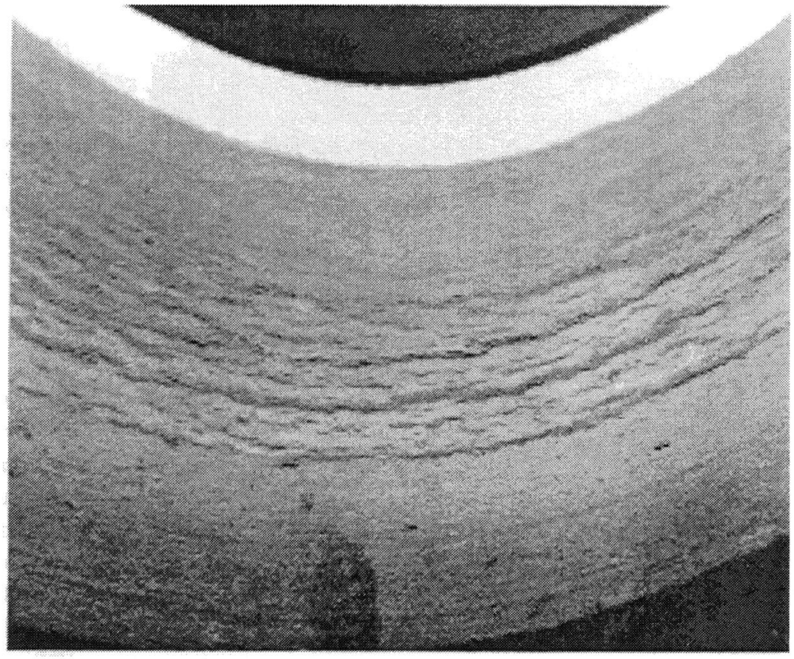

Figure 4.21(a) - Photograph of cracked inner surface of pipe elbow.

Figure 4.21(b) - Photomicrograph

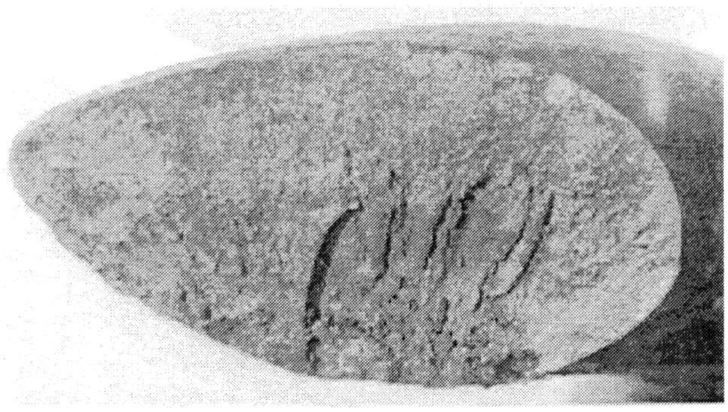

Figure 4.22(a) - Photograph of fracture surface.

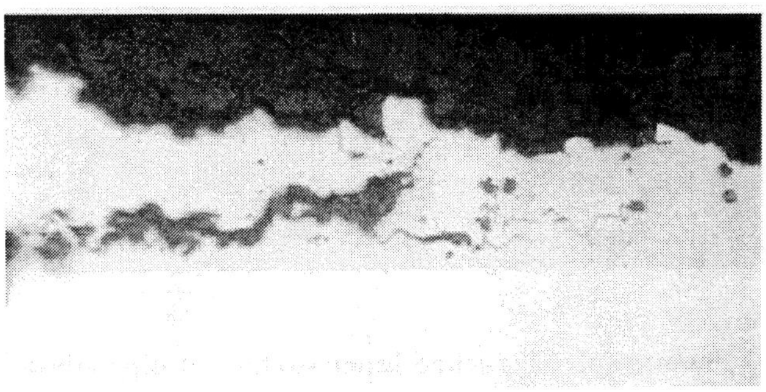

Figure 4.22(b) - Photomicrograph showing fracture profile.

CASE HISTORY 4.22

Phenomenon: Hydrogen embrittlement from corrosion-generated hydrogen
Group/ Type: low alloy steel, AISI 4140, HRC 49
Equipment: spring in a pressure relief valve
Environment: boiler feedwater
Conditions:
a) major species: boiler feedwater
b) minor species: oxygen
c) temperature: 250ºF (121°C)
d) pressure: 158 psig
e) aeration: yes

Time to failure: 13 years
Illustration: fracture surface macro, micro
Comment: This pressure relief valve spring was in piping immediately upstream of deaerater and its oxygen scavenger injection point so it was exposed to boiler feedwater which still had traces of oxygen in it. Very mild general corrosion resulted; energy dispersive x-ray analysis of the corrosion products on the fracture surface indicated that they were entirely iron oxides. The spring was so hard (HRC 49) that even the few parts per million of hydrogen generated by the mild general corrosion were enough to cause hydrogen embrittlement and the resulting intergranular fracture (see micro).
Remedy: Replace with corrosion-resistant alloy spring such as AISI 316 SS.

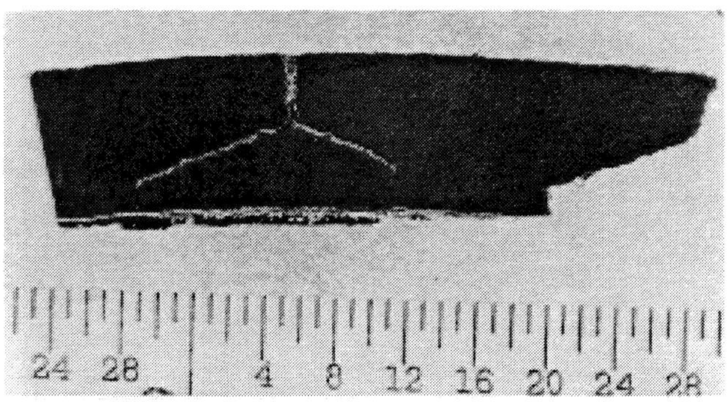

Figure 4.23 - (a)(above) Photomacrograph of external SCC of steel pipeline and (b)(below) photomicrograph showing intergranular fracture profile of external SCC failure.

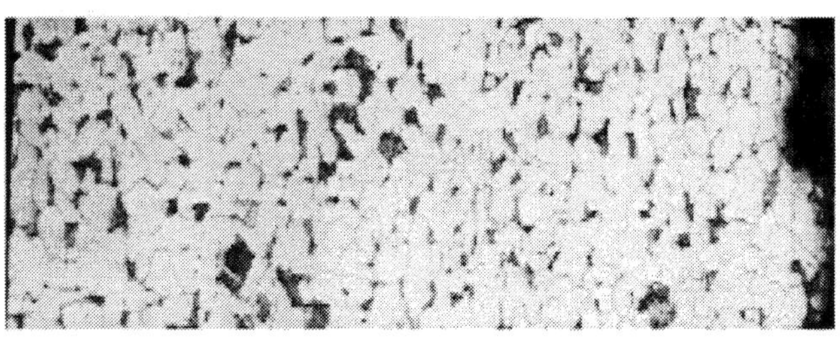

CASE HISTORY 4.23

Phenomenon: external stress-corrosion cracking
Group/Type: ferrous, carbon steel, API 5L Gr. B pipe
Equipment: crude oil pipeline
Environment: *subkha* (salt flat)
Conditions:
a) major species: water, 5% NaCl
b) minor species: carbonates, sulfates
c) temperature: Arabian Peninsula ambient
d) pressure: 1 atmosphere
e) aeration: buried to a depth of 1.5 m

Time to failure: 24 years
Illustration: macro, micro
Comment: This 14 in. diameter, 0.25 in. wall thickness crude oil line was tape-wrapped and buried in *subkha*, a sandy salt flat with a water table within 6 inches of the surface. On hydrotest the line burst at the 6 o'clock position. Metallographic examination indicated that external intergranular stress-corrosion cracking had penetrated 80% of the wall thickness (macro, micro). This type of SCC is associated with incomplete cathodic protection of steels in slightly alkaline carbonate-bicarbonate soils.
Remedy: Replace the affected areas with new pipe coated with fusion-bonded epoxy coating.

Figure 4.24(a) - Ruptured pipe section as received.

CASE HISTORY 4.24

Phenomenon: hydrogen induced cracking (HIC)
Group/Type: carbon steel, API 5L X-42 line pipe
Equipment: 24 in. diameter cross-country pipeline
Environment: sour gas/sour crude oil
Conditions:

a) major species: Arab light crude oil
b) minor species: sulfur, H_2S, 0.2 wt% water
c) temperature: Arabian ambient
d) pressure: 790 psig
e) aeration: none

Time to failure: 37 years total
Illustration: overall, metallographic cross-sections
Comment: The pipeline transmitted sour gas for 20 years, followed by 17 years service in sour crude oil. At that time a hydrostatic test at 790 psig resulted in a rupture approximately six feet long. The rupture initiated in a zone of internal corrosion and blistering centered on the 6 o'clock position, at the bottom of the pipe, where a water layer had accumulated due to improper sloping of the line during construction. Cross-sections near the rupture indicated classic hydrogen induced cracking (HIC) with the formation of large hydrogen blisters, followed by heavy step-wise cracking (SWC) as adjacent blisters grew together by shear of the ligaments between them.
Remedy: Locate all similar damaged areas with instrumented pipeline pigs, replace with HIC-resistant steel made by reducing the sulfur content and spheroidizing the inclusions with calcium treatment.

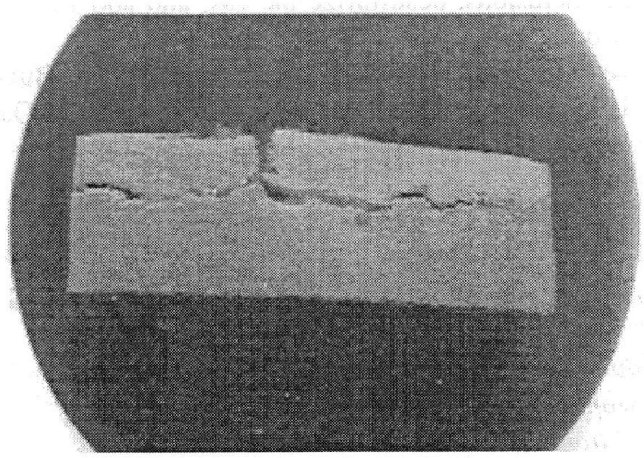

Figure 4.24(b) - Metallographic cross section showing HIC near rupture.

CASE HISTORY 4.25

Phenomenon: hydrogen blistering
Group/ Type: ferrous low alloy steel, AISI 4145H (UNS H41450), HRC 35 to 37
Equipment: tank truck cylinder
Environment: sour wellhead gas
Conditions:

(a) major species: methane
(b) minor species: H_2O, 500 ppm H_2S
(c) temperature: ambient
(d) pressure: 3,600 psi
(e) aeration: anaerobic

Time to failure: 3 weeks
Illustration: micro
Comment: Sour natural gas produces hydrogen sulfide stress cracking when stress exceeds critical levels. At lower stresses, diffusible hydrogen enters the steel then recombines at inclusions and microstructural irregularities, causing, as in this case, blisters. When such blisters begin to join up at shear steps, serious strength loss can result. This is known as stepwise cracking.
Remedy: Use calcium-treated steels with shape-controlled inclusions, desulfurize the gas, and add filming inhibitors.
Reference: B.W. Christ, "Analysis of the Pneumatic Burst of a Large Seamless Steel Pressure Vessel in Natural Gas Service," DOT final report, December, 1978.

Figure 4.25 - Micro

CASE HISTORY 4.26

Phenomenon: SCC
Group/ Type: ferrous; ASTM A516 grade 70 (UNS K02700) carbon steel
Equipment: impregnation vessel, paper mill
Environment: Kraft liquor
Conditions:

(a) major species: water + 60 g/L NaOH
(b) minor species: polysulfides, polythionates
(c) temperature: 230°F (110°C)
(d) pressure: 200 psi
(e) aeration: partial

Time to failure: 18 months
Illustration: photomicrograph
Comment: Caustic SCC in pulp and paper mills is believed to be true anodic SCC, rather than hydrogen stress cracking. Note the small secondary cracks visible along the top surface of the weld and HAZ. Primary crack propagation is along the interface between the HAZ of successive weld passes.
Remedy: Thermal relief
Reference: Battelle Northwest, Richland, WA, USA.

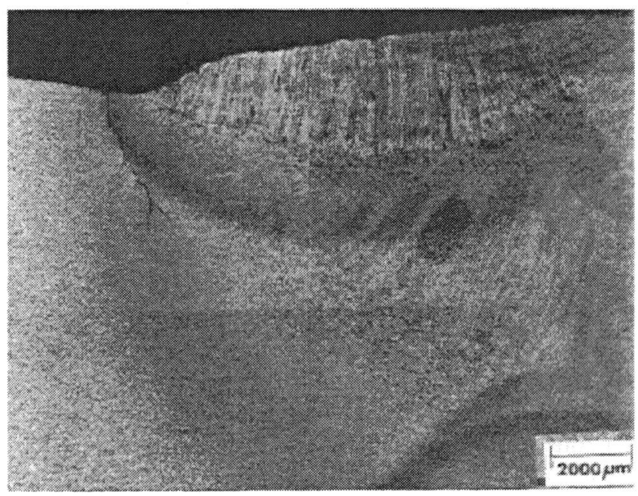

Figure 4.26 - Micro of stress corrosion crack in weld area under tensile stress.

CASE HISTORY 4.27

Phenomenon: corrosion fatigue
Group/ Type: ferrous, AISI 1012 (UNS G10120) carbon steel
Equipmment: autoclave fan
Environment: hot, moist air
Conditions:
(a) major species: air
(b) minor species: water vapor
(c) temperature: 200°F (93°C)
(d) pressure: ambient
(e) aeration: not given

Time to failure: 20 hours in service
Illustrations: none
Comments: These autoclave fan blades dispersed the effluent from a resin curing operation. After 20 hours in service at 1,000 rpm, one of the blades broke off at the hub. SEM examination revealed the poorly defined, coarse striations characteristic of corrosion fatigue. The fatigue cracks began at weld defects in the hub-to-fan fillets.
Remedy: Shot-peen the welds and coat with a coat-tar-epoxy coating to eliminate the corrosion.

CASE HISTORY 4.28

Phenomenon: SCC
Group/Type: ferrous, carbon steel
Equipment: pulp and paper digester
Environment: pulping liquor
Conditions:
(a) major species: water, 100 g/L NaOH, 35 g/L sodium sulfide
(b) minor species: sodium carbonate, sulfate, and thiosulfate
(c) temperature: 217°F (175°C)
(d) pressure: 160 psi
(e) aeration: not determined

Time to failure: ~ 10 years
Illustration: macroetched section, micro
Comment: Worldwide study of 150 pulp and paper industry digesters revealed that more than half of them had stress corrosion cracks in highly stressed structural welds at the conical transition section between the column and the top head. Cracking is predominantly intergranular and associated with weld HAZ.
Remedy: Stress-relieved vessels suffered a much lower rate of cracking (~ 10% of inspected vessels cracked as opposed to over 50% for non-stress-relieved vessels). Other preventive measures in use or proposed include stainless alloy weld overlay, thermally sprayed coatings, shot peening, and anodic protection.
Reference: D.C. Bennett, *Materials Performance*, 23, 1(1984): pp. 9-15.

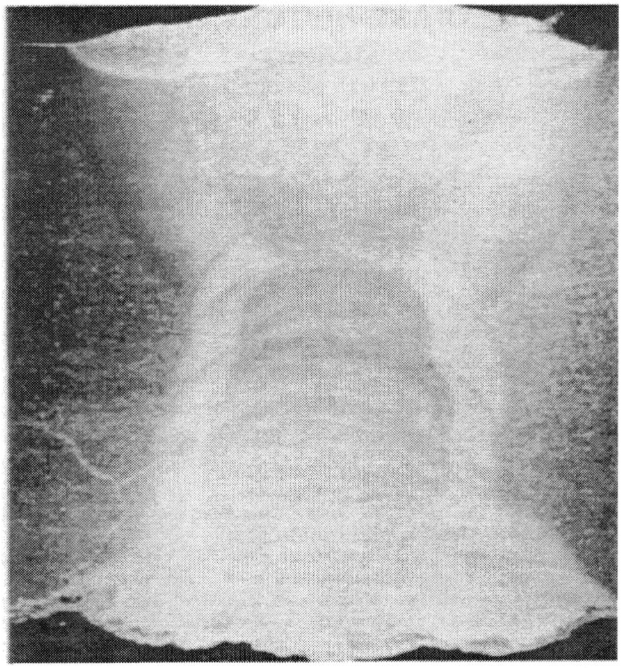

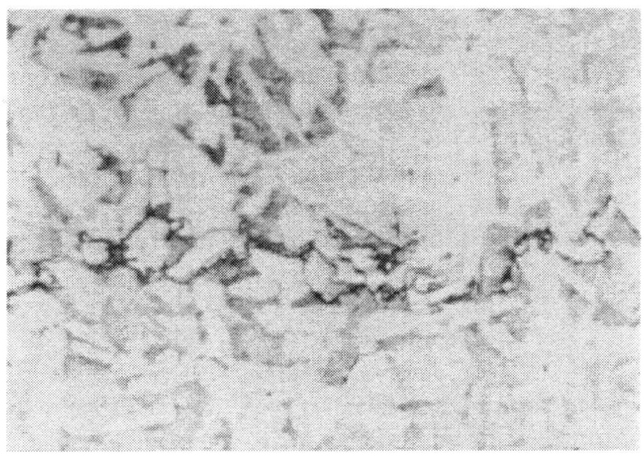

Figure 4.28 (a)(top) - Macro of continuous digester structural weld showing corrosion of weld and SCC at heat affected zone and (b) micro showing intergranular nature of the cracking in base metal.

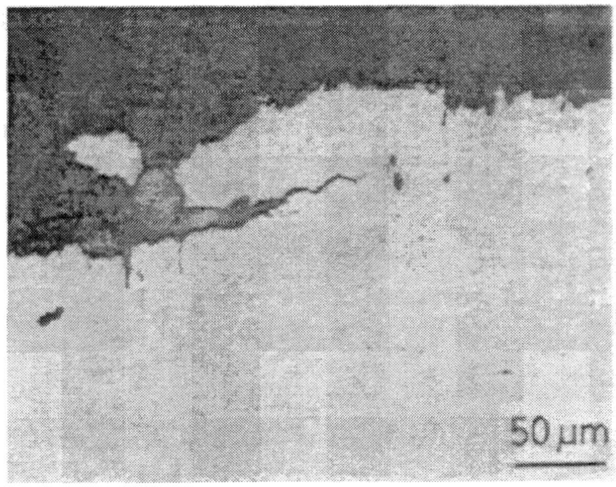

CASE HISTORY 4.29

Phenomenon: sulfide stress cracking
Group/ Type: ferrous, carbon steel, API 5AC type L-80, 99 ksi YS, tensile strength 115 ksi (HRC 22.2)
Equipment: natural gas well casing
Environment: packer fluid
Conditions:
(a) major species: 2% KCl in water
(b) minor species: inhibitor, 103 psig hydrogen sulfide, 135 psig carbon dioxide in the gas phase
(c) temperature: near ambient
(d) pressure: 4,500 psig total pressure
(e) aeration: anaerobic
Time to failure: not given
Illustration: macro, photomicrograph

Comment: Steels with hardness below HRC 22 are generally considered immune to sulfide stress cracking in sour gas wells. However, this L-80 casing joint failed by sulfide stress cracking after a packer leak allowed sour gas to contaminate the packer fluid between the casing and tubing. The packer leak produced excess pressure on the casing, which was estimated to have produced axial tensile stresses approaching the yield stress. Although the average hardness was only HRC 22.2, examination showed that a few banded regions in the micro-structure showed micro-hardness as high as KHN 275 (HRC 24.5).
Remedy: Specify lower hardness limits for L-80 casing for sour service.
Reference: S.W. Ciaraldi, "Material Failures in Sour Gas Service," CORROSION/85, paper no. 217 (Houston, TX: NACE, 1985)

CASE HISTORY 4.30

Phenomenon: sulfide stress cracking
Group/ Type: ferrous, ASTM A234 (UNS K03006) WPB carbon steel, 68 ksi YS, 83.5 ksi tensile strength
Equipment: reducer in gas gathering pipeline
Environment: hydrocarbon condensate + sour water
Conditions:
(a) major species: light hydrocarbons, water
(b) minor species: inhibitor, 165 psig H_2S, 57 psig CO_2
(c) temperature: 109°F (43°C)
(d) pressure: 1,100 psig
(e) aeration: anaerobic
Time to failure: not given
Illustration: macro; micro with microhardness readings

Comment: Steels with less than 80 ksi YS are normally considered immune to sulfide stress cracking. However, this carbon steel reducer fitting had apparently been given a final sizing press at temperatures below the stress-relieving temperature. As a result, both inner and outer surfaces had thin bands with hardness much higher than normal (KHN 285 to 315). This particular fitting was subjected to unusually high stresses caused by nearby ground settling. Once sulfide stress cracks began in the cold-worked, high hardness zones on the surface, the cracks propagated through the lower hardness metal below to cause leakage.
Remedy: Specify a final normalization treatment for any fitting to be used in sour service.
Reference: S.W. Ciaraldi, "Materials Failures in Sour Gas Service," CORROSION/85, paper no. 217 (NACE, 1985).

CASE HISTORY 4.31

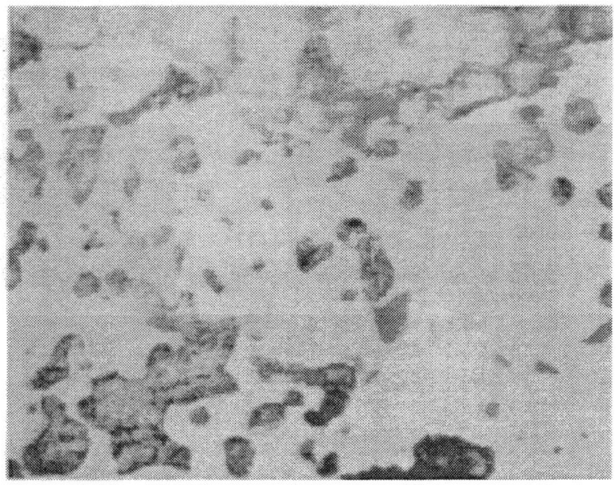

Figure 4.31 - (a) Crack of mild steel tuyere pipe exposed to molten copper matte and slag and (b) dissolution of freed grains in molten copper.

Phenomenon: liquid metal cracking
Group/ Type: ferrous, carbon steel
Equipment: copper converter tuyere
Environment: copper ore conversion
Conditions:
(a) major species: molten copper matte, slag
(b) minor species: air
(c) temperature: unknown (converter refractory is maintained at 2,200°F (1,204°C). Tuyeres are cooled by the rush of blowing air)
(d) pressure: atmospheric
(e) aeration: shielded by refractory
Time to failure: 72 hours

Illustration:
Comment: Reaction air blown through the tuyeres cools them somewhat below the 2,200°F (1,204°C) maintained inside the converter. However, copper penetration of the refractory can expose the external tuyere surface to molten copper matte (a mixture of copper, copper sulfide, iron, iron sulfide, and gangue), which can lead to rapid liquid metal cracking from the outside in exposed areas.
Remedy: Prevent copper penetration of the refractory or use copper-nickel alloy tuyeres.
Reference: S.L. Pohlman, R.V. Olson, "Corrosion and Material Problems in the Copper Production Industry," CORROSION/84, paper No. 229 (NACE, 1984).

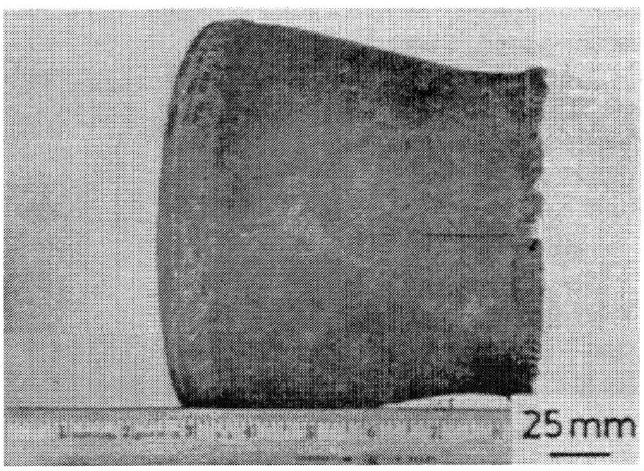

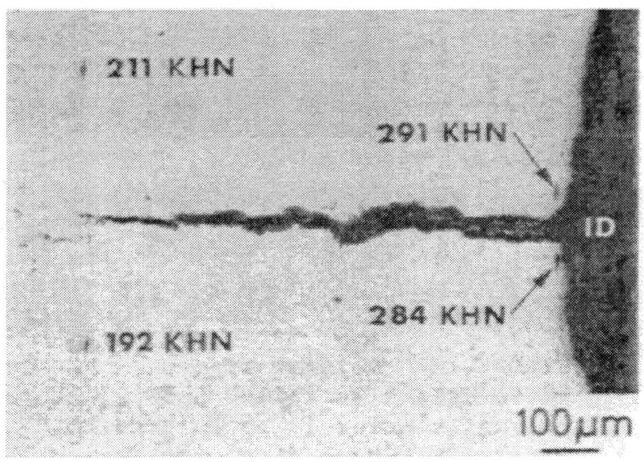

Figure 4.30 - Macro and micro with microhardness readings

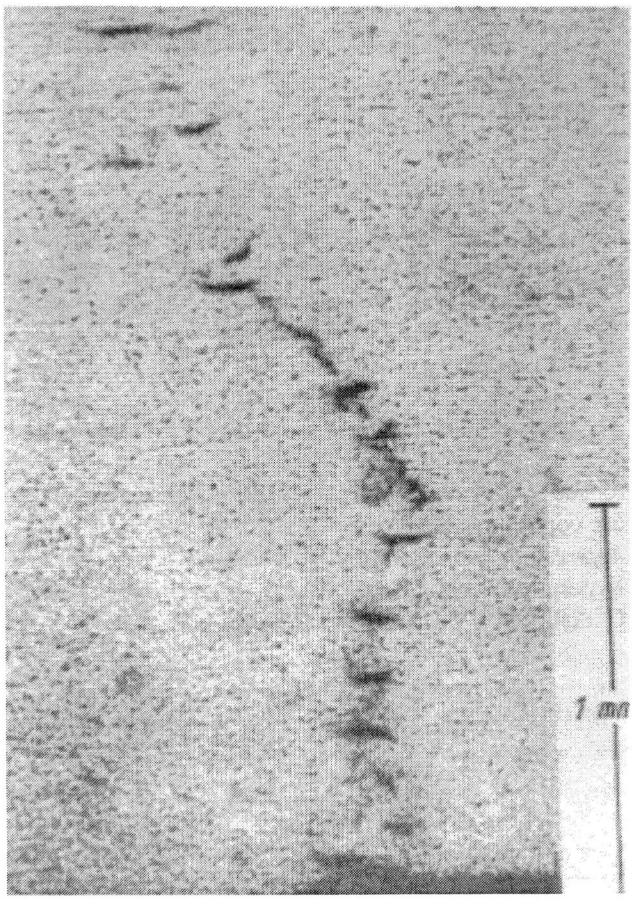

CASE HISTORY 4.32

Phenomenon: stress-oriented, hydrogen-induced cracking (SOHIC)

Group/ Type: ferrous/RRSt 43.7 (ERW) DIN 17172 carbon steel, C = 0.22%, S = 0.018%, YS = 313 N/mm^2 (45 ksi) YS, UTS = 452 N/mm^2 (65 ksi)

Equipment: gas pipeline

Environment: dehydrated natural gas

Conditions:

(a) major species: methane, 23% H_2S, 8.5% CO_2

(b) minor species: 30 to 80 mg/m^3 water in dried gas, 1,200 mg/L chlorides in liquid, traces of triethylene glycol

(c) temperature: ambient

(d) pressure: 88 bar (1,276 psi)

(e) aeration: 100 to 200 ppm oxygen in production-strip gas mix

Time to failure: 7 years

Illustrations: macrograph, photomicrograph

Comment: Hydrogen blisters connecting at stress-oriented radial shear steps linked up to cause the rupture of a sour gas pipeline, with fatality. This line handles only gas dehydrated with triethylene glycol so the source of the nascent hydrogen was not readily apparent. Large quantities of elemental sulfur were found in the line. The investigators concluded that the high oxygen levels introduced at the glycol stripping plant were oxidizing the hydrogen sulfide to water and elemental sulfur. The water thus produced nullified the dehydration process and allowed corrosion to proceed in the line. Corrosion-generated hydrogen then caused blistering and HIC, leading to the rupture.

Remedy: First, oxygen in the glycol stripper gas was eliminated. Second, monoethanolamine was injected at the rate of 1.5 mg/standard m^3 as a corrosion inhibitor. Last, desulfurized, inclusion shape-controlled steels were specified for new construction.

Reference: W. Bruckhoff, O. Geier, K. Hofbauer, "Rupture of a Sour Gas Line due to Stress-Oriented Hydrogen Induced Cracking: Failure Analysis, Experimental Results and Corrosion Prevention," CORROSION/85, paper no. 389 (Houston, TX: NACE, 1985).

CASE HISTORY 4.33

Phenomenon: hydrogen-assisted cracking
Group/ Type: ferrous/low alloy steel (0.44 carbon. 0.92 manganese, 0.98 chromium, 0.20 molybdenum, UTS = 170 ksi
Equipment: natural gas tube trailer
Conditions:
(a) major species: methane
(b) minor species: 800 ppm H_2O, 550 ppm H_2S
(c) temperature: ambient
(d) pressure: 2,500 psi
(e) aeration: anaerobic
Time to failure: 3 weeks total (less than 1 hour during this filling)
Illustration: as-received, micro
Comment: A tube trailer cylinder burst while being filled with wet, sour natural gas. Burst pressure was estimated at 2,500 psi. Although HRC averaged 36, a few microscopic hard spots showed hardness averaged as high as HRC 45. At the fracture initiation site, classic intergranular fracture surface features were observed, indicating failure by sulfide stress cracking, a form of hydrogen-assisted cracking.
Remedy: Use aluminum-alloy cylinders or lower the hardness of steel cylinders below HRC 22, or remove water and H_2S before over-the-highway transport.
Reference: B.W. Christ, "Analysis of the Pneumatic Burst of a Large Seamless Steel Pressure Vessel in Natural Gas Service," US DOT final report, Dec. 1978.

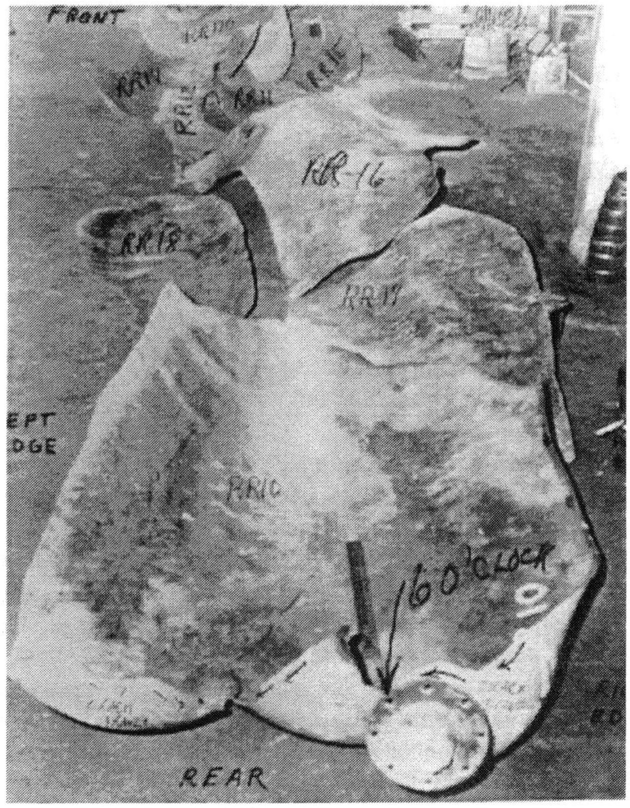

CASE HISTORY 4.34

Phenomenon: stress-corrosion cracking
Group/ Type: carbon steel, API 5L Gr. B piping
Equipment: gas treating plant piping
Environment: propane
Conditions:

a) major species: propane
b) minor species: traces of 17% NaOH, sulfur, mercaptans
c) temperature: ambient, 118ºF (48°C) maximum
d) pressure: 330 psig
e) aeration: unknown

Time to failure: 10 years
Illustration: photomicrograph
Comment: SCC of carbon steel in NaOH service is normally considered to show a temperature threshold, as outlined by the NACE Caustic Soda Service Chart, below which SCC does not occur. However, in this case, as-welded plant piping developed cracking even though the service temperature was below the threshold for SCC. NaOH carryover was postulated due to carryover from an upstream gas sweetening unit being operated beyond design flow rates, and traces of NaOH were detected in the fracture surfaces. Laboratory slow-strain rate testing established that the plant caustic, high in sulfur and mercaptans, showed a temperature threshold for SCC well below that of reagent grade solutions.
Remedy: Stress-relieve welded piping for NaOH service regardless of service temperature.
Contributor: S.P. Cox, Saudi Aramco, Dhahran, Saudi Arabia

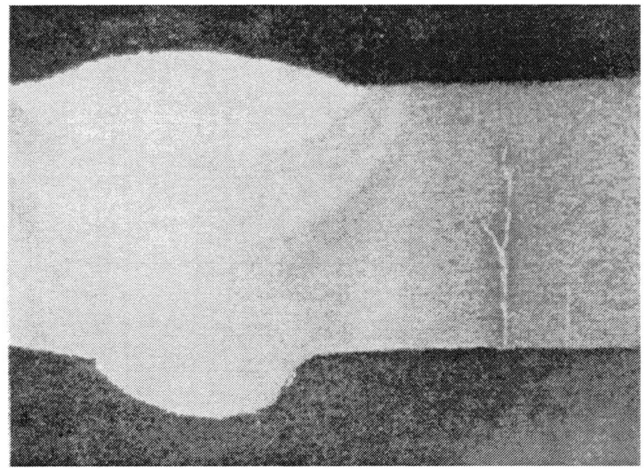

Figure 4.34 - Micro of SCC crack.

CASE HISTORY 4.35

Phenomenon: sulfide stress cracking
Group/ Type: ferrous, 5 Chromium low alloy steel (UNS K41545)
Equipment: 2 in. diameter gas well piping
Environment: sour natural gas
Conditions:

a) major species: methane, condensed light hydrocarbons
b) minor species: 4% H_2S, 2% CO_2
c) temperature: Arabian ambient (50-130ºF [10-54°C])
d) pressure: 1,100 psig
e) aeration: anaerobic

Time to failure: 2 years
Illustrations: macro, micro (pages 66 and 67)
Comment: This 5Cr steel pipe elbow was welded into a carbon steel line by mistake. In the weld heat-affected zone the low alloy steel elbow developed hardnesses of VHN 435 to 486. After two years service, a leak developed around 150 degrees of the circumference of the weld. Metallography indicated that the crack was branched, intergranular and propagated solely through the hardened heat-affected zone. The only surprising thing about this sulfide stress cracking failure is that it lasted as long as it did. Steels of this hardness would normally be expected to fail in hours in wet sour service. However, this particular well gave gas with a low water content and some natural organic inhibitors, which slowed the time to failure.
Remedy: Eliminate all 5Cr steel items from wet sour service with a positive materials identification program (PMI) using chemical spot tests or x-ray fluorescence analysis.

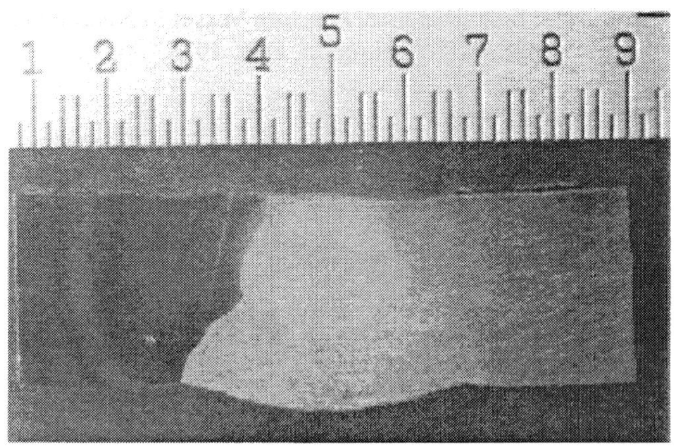

Figure 4.35(a) - Photomacrograph of mounted cross-section

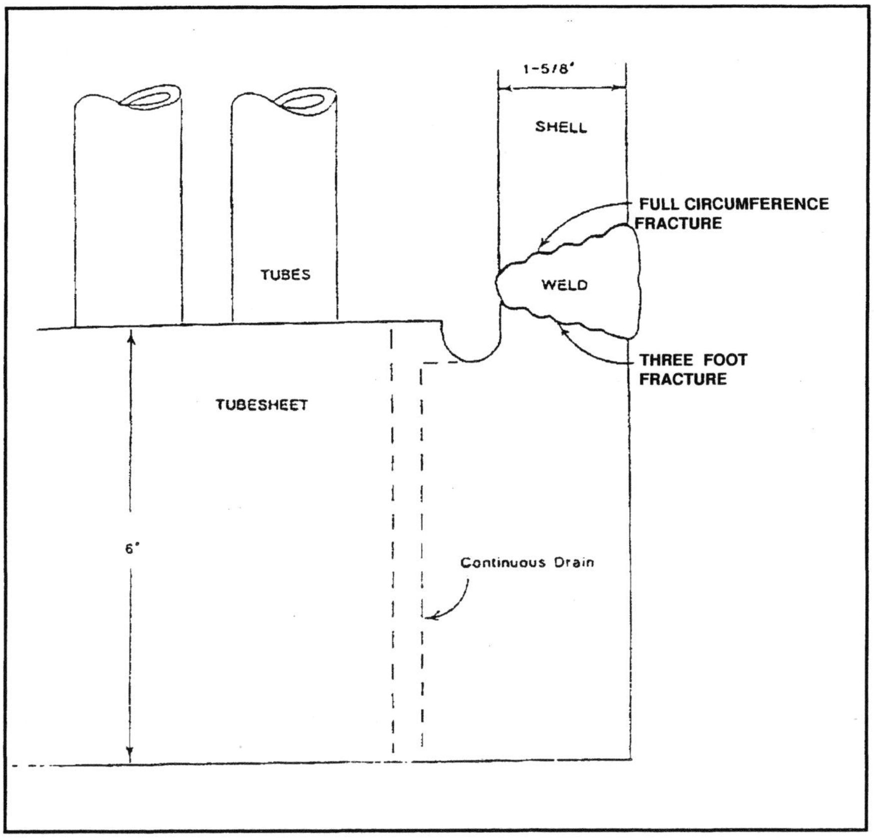

Figure 4.36 - Schematic of dissimilar metals weld, tubesheet-to-shell joint.

CASE HISTORY 4.36

Phenomenon: sulfide stress-cracking

Group/ Type: ferrous, 3½ % Ni steel welded with 18-8W SS, 1,100ºF (598°C) stress relief

Equipment: 78-in. diameter vertical heat exchanger

Environment: hydrogen-based process gas

Conditions:

a) major species: 60% H_2, 31.6% CO_2, 7% CO, water saturated

b) minor species: 0.2% CH_3, 0.9% H_2S, "several ppm" HCN and NH_3

c) temperature: -30 to +30ºC (-34.4 to -1.1°C)

d) pressure: 950 psig

e) aeration: anaerobic

Time to failure: 5 years

Illustration: schematic of tubesheet-to-shell joint, fracture

Comment: This vertical heat exchanger had the shell welded to the tubesheet using a tungsten-modified 18-8 weld metal to match the thermal expansion characteristics of the 3½ % Ni steel sheel. After some years in service the heat exchanger failed catastrophically, rupturing completely around the circumference at the fusion line between the 3½ % Ni steel sheel and the 18-8W weld metal. Investigation showed that a thin band of metal at the fusion line had an air-hardening composition high in chromium, nickel, and carbon, which readily formed martensite of KHN500g 310 hardness on cooling from the welding temperatures. This thin mixed zone resisted tempering at the 1,100ºF stress relief temperature and was extremely sensitive to sulfide stress cracking.

Remedy: Eliminate the 18-8W weld metal, and fabricate with a filler metal matching the 3½ % Ni steel base metal.

Reference: B.D. Craig and R.B. Setterlund, "Catastrophic SSC Failure of a Dissimilar Metal Weld in a High Pressure Hydrogen Vessel," CORROSION/91, paper no. 317 (Houston, TX: NACE, 1991)

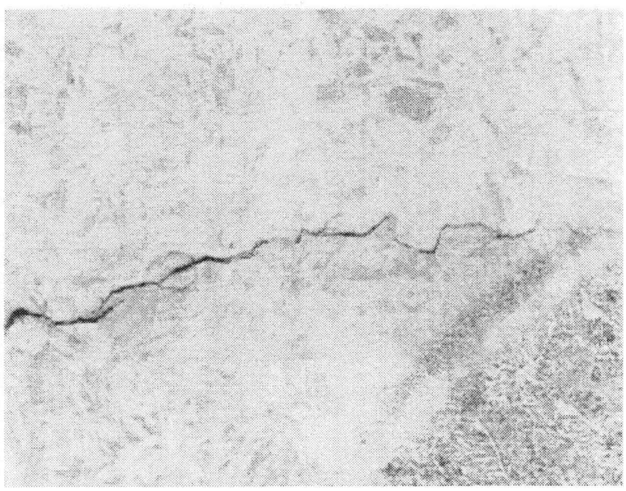

Figure 4.35(b) - Photomicrograph of crack in mounted section in 4.35(a).

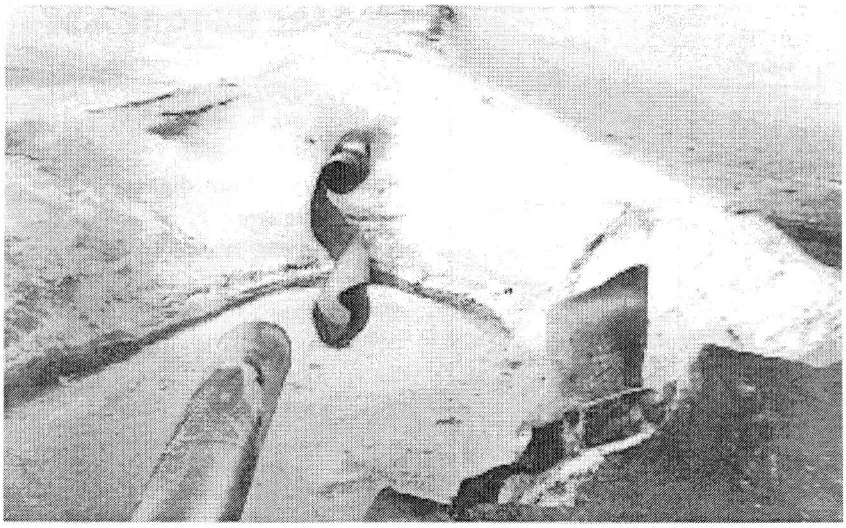

Figure 4.37(a) - (left) Pipe rupture which occurred after six weeks in service.

Figure 4.37(b) - (below) Photomicrograph of SOHIC cracks near spiral weld seam.

CASE HISTORY 4.37

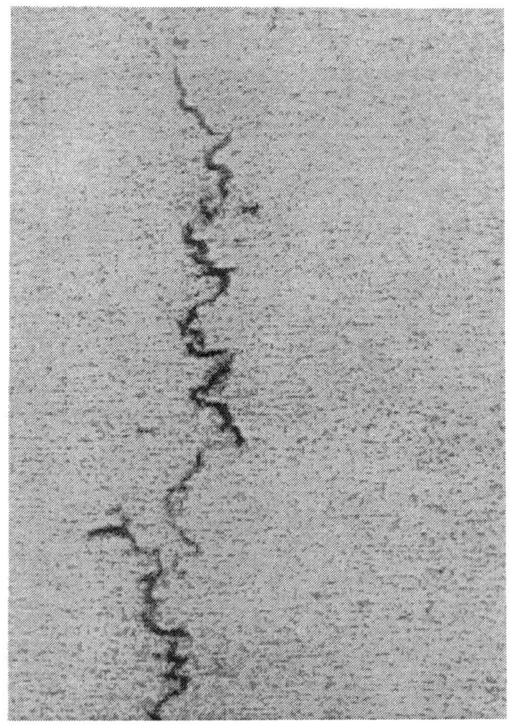

Phenomenon: stress-oriented hydrogen-induced cracking
Group/Type: ferrous, carbon steel, API 5L Gr. X-42, Y.S. 64 ksi, UTS 74.5 ksi, hardness HV 174, composition C = 0.14%, S = 0.024%, P = 0.019%, Si = 0.25%, Mn = 0.99%, Cu = 0.04%, Al = 0.043%
Equipment: cross-country gas transmission pipeline
Environment: wet sour gas
Conditions:

a) major species: methane
b) minor species: 3.4% H_2S, 8.8% CO_2
c) temperature: 27-38°C (80-100°F)
d) pressure: 350 psig
e) aeration: anaerobic

Time to first failure: 10 days
Illustration: photograph, photomicrograph
Comments: This 24 in. diameter, 0.25 in. wall thickness gas transmission line suffered three ruptures in service shortly after commissioning, the last of which caused an explosion, fire, and major spill. Subsequent repairs and refurbishing revealed 20 additional zones which burst on hydrostatic testing at 700 psig. The fractures propagated along the heat-affected zone of the spiral seam welds. Weld heat-affected zone hardness was low, HV 148-174. The pipe manufacturing process produced high residual stresses and large weld crowns. The pipe was made of controlled-rolled skelp, which was particularly sensitive to hydrogen-induced cracking and stepwise cracking. Water collecting in low spots in the line caused corrosion, HIC and SWC. The high residual stresses and the stress raisers caused by the weld crown geometry resulted in SOHIC (stress-oriented hydrogen induced cracking) near the spiral weld seams.
Remedy: Reduce the working pressure to limit the hoop stress to < 25% of the specified minimum yield stress, and inject corrosion inhibitor at four times the normal rate. This line was operated in this fashion for two years before being switched to sweet gas service. New construction for wet sour gas transmission was specified as HIC-resistant steel with calcium additions for inclusion shape control.
Contributor: E.M. Moore Jr., Saudi Aramco, Dhahran, Saudi Arabia.

CASE HISTORY 4.38

Phenomenon: external SCC
Group/ Type: austenitic stainless steel, AISI 304 (UNS S30400)
Equipment: reboiler flange
Environment: marine industrial atmosphere
Conditions:
(a) major species: air
(b) minor species: polyvinyl chloride resin fines
(c) temperature: 400°F (204°C)
(d) pressure: atmospheric
(e) aeration: total
Time to failure: 30 years
Illustration: photomicrograph

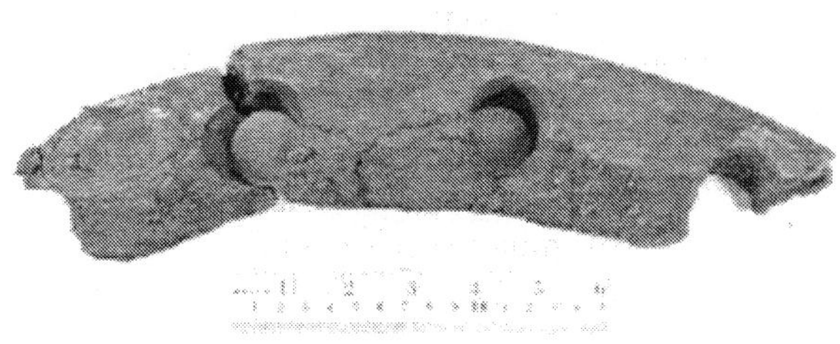

Comment: Polyvinyl chloride dust may hydrolyze at temperatures above 80°C, giving off HCl. As a result, external SCC is common in PVC manufacturing units since it is frequently difficult to keep the equipment completely free of PVC resin fines. This 304 SS flange was installed in a steam heater for a distillation column in a polyvinyl chloride plant. No degradation occurred on the steam side. However, after 30 years in service the flange was so riddled with external stress corrosion cracks due to the breakdown of PVC dust in the area that pieces could be broken off by hand.
Remedy: Replace with clad steel.

CASE HISTORY 4.39

Phenomenon: external SCC
Group/ Type: stainless steel, AISI 304L (UNS S30403)
Equipment: pressure vessel
Environment: hydrotest water + ambient air
Conditions:
(a) major species: city water, air
(b) minor species: chlorides (150 ppm)
(c) temperature: cryogenic (steady state). Unknown during fire.
(d) pressure: ambient
(e) aeration: saturated
Time to failure: four years total service life; ~3 to 6 hours of high-temperature exposure during fire.
Illustration: cross-section macro and micro

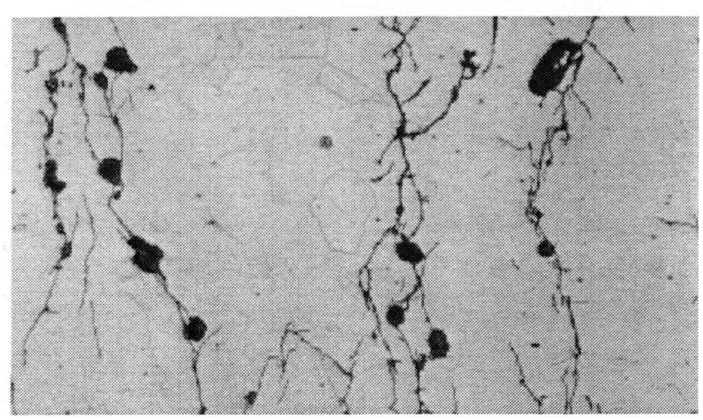

Comment: This AISI 304L SS pressure vessel was used for several years in cryogenic service in an oxygen plant. All the nozzle reinforcement pads had pressure fittings threated into them so the reinforcement pad welds could be hydrostatically tested with city water. However, once hydrotested, it is imposssble to clean and dry the space between the vessel wall and the pad. In service, the pressure-fitting ports were left open, and condensation on the cold vessel walls kept this area wet. Since the vessel normally operated at cryogenic temperatues this caused no problems for several years. However, am accidental fire in the area resulted ina brief exposure to temperatures well above normal. On examination after the fire, the vessel showed extensive leakage and SCC beaneath the reinforcing pad. Note on the photomicrographs that at the end of some of the crack branches, a spherical area of pitting is observed. Apparently when the stresses that drive SCC have been relieved, the crack tip is strongly anodic to the crack sides. Thus, pitting proceeds even though crack propagation does not.
Remedy: Hydrotest with air or nitrogren, then seal off the test ports with threaded fittings.

CASE HISTORY 4.40

Phenomenon: external SCC
Group/ Type: AISI 316L SS (UNS S31603)
Equipment: autoclave
Environment: calcium sulfate + calcium chloride + steam
Conditions:
(a) major species: $CaSO_4$, $CaCl_2$, H_2O
(b) minor species: O_2
(c) temperature: 300°F (149°C)
(d) pressure: 34 psig (process) ambient out side
(e) aeration:
Time to failure: 1 year
Illustration: macro

Comment: In this autoclave, calcium suflate was saturated with calcium chloride, then heated for 4 hours in 34 psig saturated steam. The process was essentially deaerated and, despite the high chloride content, no cracking was observed om the inside surfaces. However, the locking rings allowed some steam leakage. The process steam would condense on the locking ring and locking lugs, leaving hot water droplets in $CaCl_2$. Oxygen pickup from ambient air made this solution extremely aggressive. After one year of operation, the locking rings and lugs were covered with pitting and inoperable because of extensive SCC.
Remedy: Use locking rings and lugs weld overlaid with a SCC-resistant nickel-based alloy.

Figure 4.41(a)

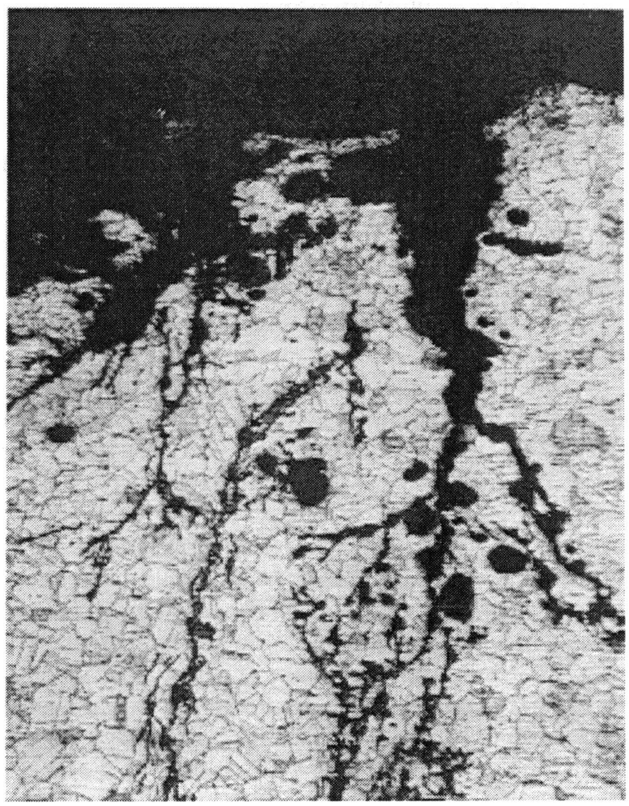

Figure 4.41(b)

CASE HISTORY 4.41

Phenomenon: SCC
Group/ Type: AISI 304 SS (UNS S30400)
Equipment: autoclave cooling coils
Environment: cooling water inside, ambient air outside
Conditions:
- (a) major species: water (inside), air (outside)
- (b) minor species: chlorides (50 ppm in water) oxygen
- (c) temperature: cyclic operation, 60° to 300°F (15.6° to 149°C)
- (d) pressure: 125 psi
- (e) aeration: saturated

Time to failure: 4 years
Illustration: macro, micro (page 70)
Comment: This vessel had a spiral half-pipe AISI 304 (UNS S30400) cooling jacket TIG welded to its outer surface. Mineral wool insulation was held in place with steel chicken wire. Process temperatures were controlled by cold [50°F (10°C)] water followed by steam condensate [360°F (182°C)] in the cooling coils. After four years, numerous external stress corrosion cracks were observed under the insulation, many at points where the steel chicken wire had left rust stains on the stainless. Examination showed a few internal stress cracks on the water side as well, although the cooling water was inhibited with ~ 15 ppm orthophosphates. The alternating cold and hot water caused condensation of atmospheric moisture under the insulation followed by evaporation and concentration of atmospheric chlorides and SCC. At the same time, the alternate hot and cold water led to serious water salt deposits on the inside of the coil. Water-side stress corrosion cracks began at cold-worked areas on the inside surface where chlorides had concentrated beneath these deposits.
Remedy: Replace with steel vessels internally clad with stainless with thicker steel cooling coils. Alternatively, use a silicone paint on the outer surface of the stainless and use chloride-free water for both heating and cooling.

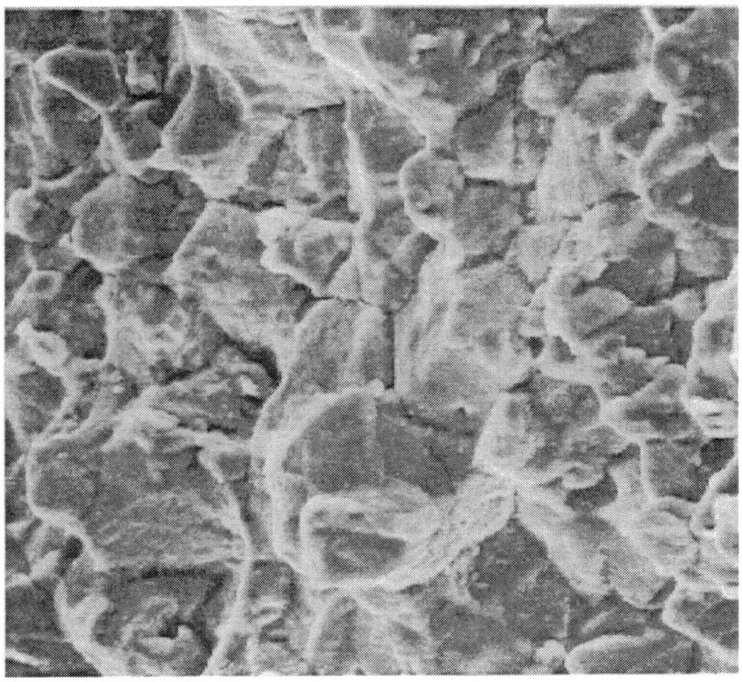

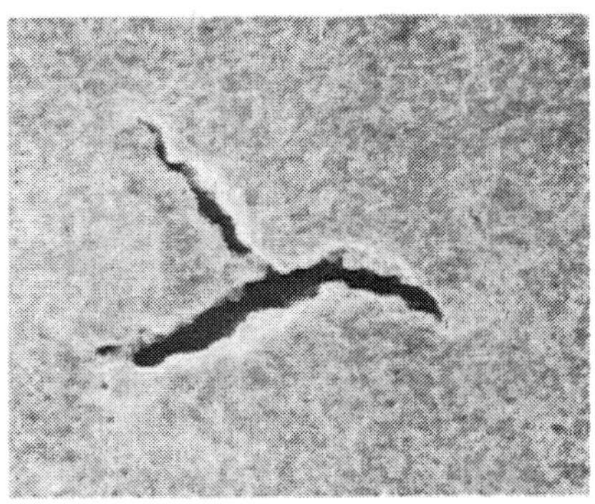

Figure 4.42 (a) and (b) - SEM fractographs.

CASE HISTORY 4.42

Phenomenon: liquid metal embrittlement
Group/ Type: SS, alloy 21Cr-6Ni-9Mn (UNS S21900)
Equipment: fire extinguisher bottle
Environment: shop fabrication
Conditions:
- (a) major species:
- (b) minor species: copper
- (c) temperature: 2,000+°F (1,093+°C)
- (d) pressure: atmospheric
- (e) aeration: inert-gas shielded

Time to failure: immediate
Illustration: SEM fractographs
Comment: Copper chill bars were clamped to a stainless fire extinguisher bottle to reduce sensitization during TIG-welding of attachment clips. Upon hydrostatic testing, some small cracks were observed near the welds. Intergranular deposits were observed on metallography, which proved to be high in copper. Therefore, the cracks were attributed to liquid metal cracking during welding.
Remedy: Discontinue using copper fixturing, and prevent sensitization by adjusting welding heat input.

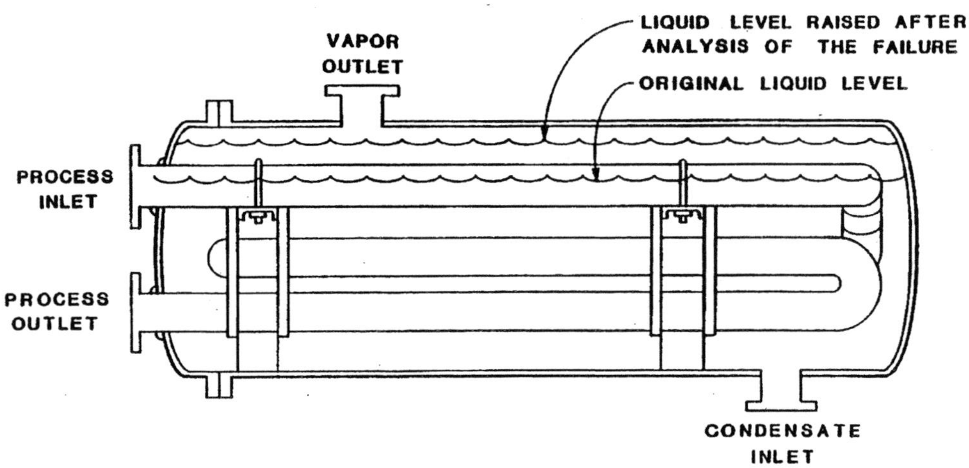

CASE HISTORY 4.43

Phenomenon: SCC
Group/Type: AISI 316L SS (UNS S31603)
Equipment: reactor coil
Environment: steam condensate
Conditions:
(a) major species: water
(b) minor species: intermittent chloride leakage to 13 ppm
(c) temperature: 374°F (190°C)
(d) pressure: 20 psig
(e) aeration: not determined
Time to failure: 8 months
Illustration: macro
Comment: This reactor was temperature-controlled with steam condensate on the shell side. Normal chloride content of the condensate was zero, but occasional condenser leaks resulted in excursions of up to 13 ppm Cl-. By chance, the liquid level on the shell side was maintained in the center of the top row of tubes. Flashing and boiling of the condensate concentrated the low levels of chlorides to levels high enough to crack the top row of tubes. All other tubes, which were fully immersed in the condensate, suffered no damage.
Remedy: Raising the liquid level in the shell to cover the top row of tubes gave four years of trouble-free service before unit shutdown.

CASE HISTORY 4.44

Phenomenon: corrosion fatigue
Group/ Type: AISI 316 SS (UNS S31600)
Equipment: valve and piping
Environment: industrial/marine atmosphere
Conditions:
(a) major species: air
(b) minor species: leakage from nearby salt slurry pump
(c) temperature: 302°F (150°C)
(d) pressure: ambient
(e) aeration: total
Time to failure: 3 months
Illustration: macro, fracture surface (page 73)
Comments: Corrosion fatigue is perhaps underrated as a source of failures in chemical plants. This failure occurred because a relatively large 11 lb (5kg) valve was hung off a ¾-in. SS nipple in an area where vibration from a nearby salt slurry pump was significant. Leakage of salt slurry from the pump added to the aggressiveness of the environment. The fracture surface showed the characteristic smooth, beach-marked appearance of a reversed-bending fatigue failure. The microstructure at the crack initiation site showed a transgranular non-branching crack profile typical of fatigue, with no evidence of IGA or SCC.
Remedy: Replace with a weldolet, thicker nipple, and smaller valve, and add bracing to lower stress on the nipple.

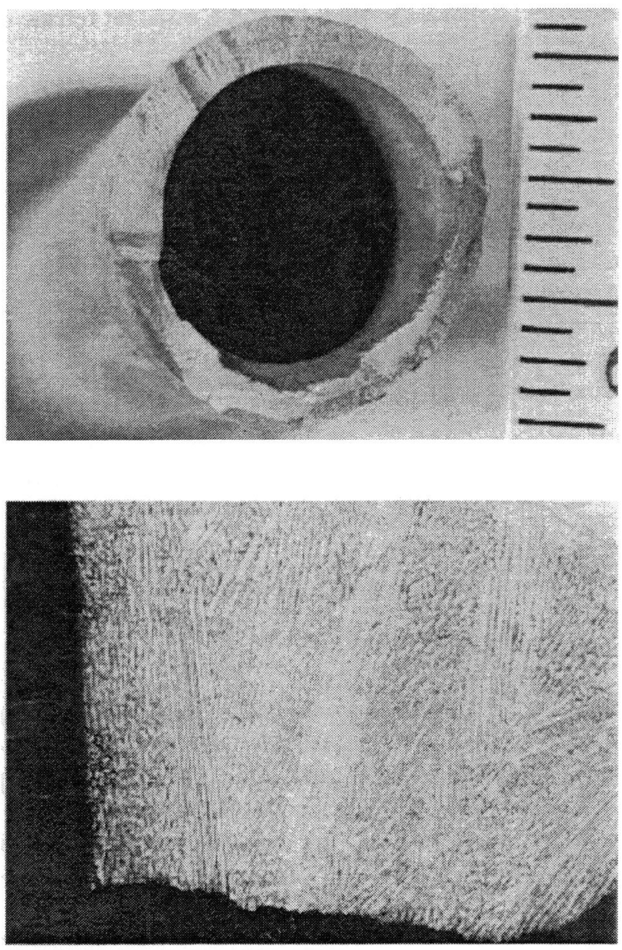

Figure 4.44 - Fracture surface macro and micro.

CASE HISTORY 4.45

Phenomenon: corrosion fatigue
Group/ Type: AISI 316 SS (UNS S31600)
Equipment: vacuum pump shaft
Environment:
sour refinery gas
Conditions:
(a) major species: SO_2, H_2O vapor, pH 1.8
(b) minor species: H_2S, 0.03 ppm Cl-
(c) temperature: 100°F (38°C)
(d) pressure: vacuum
(e) aeration: anaerobic

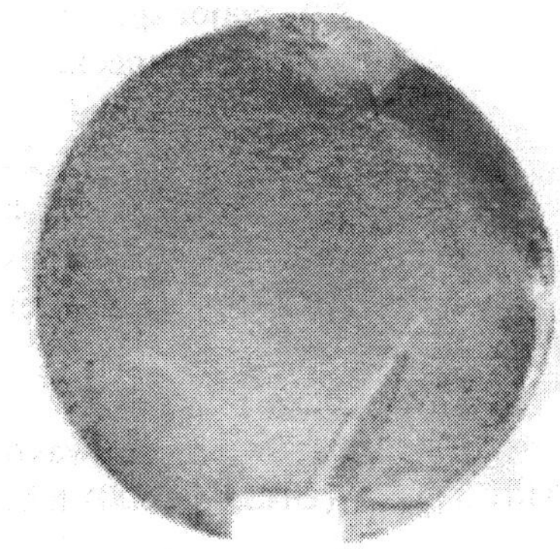

Time to failure: 6 days at 1,750 rpm
Illustration: fracture surface macro, micro
Comment: Corrosion fatigue cracks developed quickly at the sharp edges of a keyway in this pump shaft. Fracture mode (micro) is predominantly transgranular with occasional short crack branches. "Endurance limits" often are not observed in corrosive environments. Therefore, rotating equipment for corrosive service is particularly sensitive to stress raisers at geometrical details such as this keyway.
Remedy: Redesign the keyway to provide a generous radius, shot-peen, or replace with a stronger alloy, such as 17-4 PH SS.

Figure 4.45(b) - Fracture surface micro, 100 X

CASE HISTORY 4.46

Phenomenon: SCC
Group/Type: AISI 316 SS (UNS S31600)
Equipment: feed/tails intercooler
Environment: scrubber water
Conditions:

(a) major species: water
(b) minor species: 700 ppm Cl-, 50 ppm organic acid
(c) temperature: 248°F (120°C) tubeside, 122°F (50°C) shell-side outlet
(d) pressure: not given
(e) aeration: oxygen-saturated

Time to failure: 1 year
Illustration: none
Comment: This intercooler was originally installed with AISI 316 SS (UNS S31600) tubes and tubesheet. The shell was nickel-copper alloy 400. After 4 years in service, an inspection revealed severe general corrosion of the shell. However, despite the acid chlorides on the shell side, the AISI 316 (UNS S31600) stainless tubes were not damaged. A new AISI 316 (UNS S31600) stainless shell was installed. Within one year, one-third of the tubes were leaking at the U-bends because of classic transgranular SCC. Investigation suggested that the active corrosion of the Monel shell provided a measure of cathodic protection to the stainless tubes. Copper ions released into solution also apparently shifted the redox potential of the tubes masked from direct cathodic protection. Therefore, when the Monel was removed, rapid SCC of the stainless tubes resulted.
Remedy: Replace the tubes with SCC resistant ferritic or duplex SS.

CASE HISTORY 4.47

Phenomenon: SCC
Group/ Type: AISI 304 SS (S30400)
Equipment: nuclear fuel rod cladding
Environment: primary pressurized water reactor coolant
Conditions:

(a) major species: water
(b) minor species: H_2, boric acid, LiOH, O_2
(c) temperature: 532 to 578°F (277.7 to 303.3°C)
(d) pressure: not given
(e) aeration: not given

Time to failure: 2 years
Illustration: micro
Comment: A fuel pellet chip lodged between the outer surface of the pellet and the inner surface of the cladding. A stress corrosion crack developed in the cladding on the outer surface opposite to this chip.
Remedy: Reduce fuel pellet chipping and increase gap between pellet outer diameter and cladding inner diameter.
Contributor: H. Ocken, Electric Power Research Institute

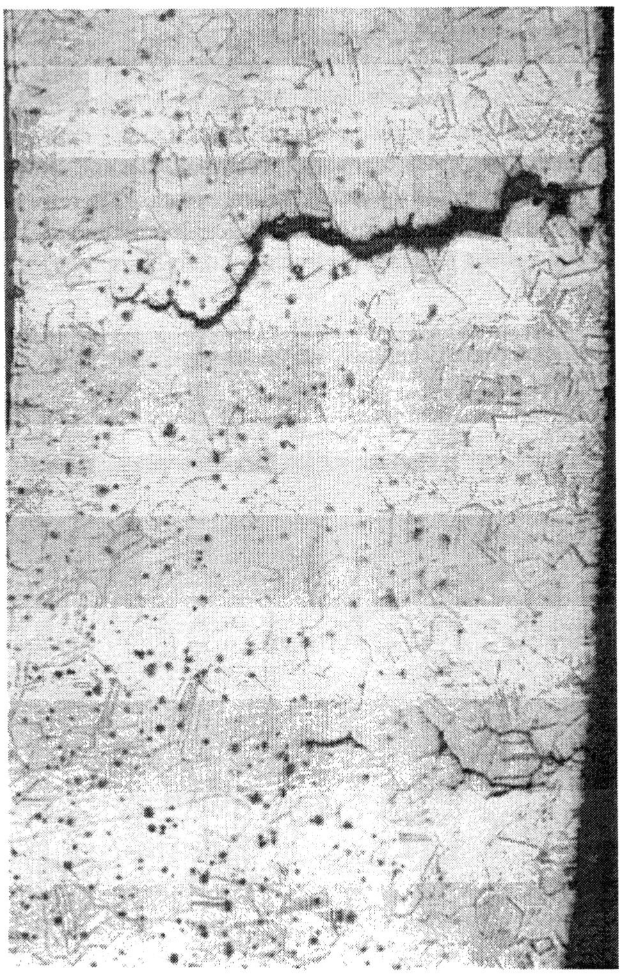

Figure 4.47 - (left) Micro

CASE HISTORY 4.48

Phenomenon: stress corrosion cracking
Group/ Type: SS, proprietary austenitic non-metallic drill collar alloy, 18% Cr, 18% Mn, 0.15% C
Equipment: drill collar
Environment: sea water
Conditions:
(a) major species: water, 3.5% NaCl
(b) minor species: Ca, Ba, Sr, etc.
(c) temperature: ambient
(d) pressure: hydrostatic pressure proportional to depth
(e) aeration: air-saturating
Time to failure: 3 years
Illustration: photograph, photomicrograph
Comment: SCC is not normally considered a problem at ambient temperatures. However, non-magnetic drill collars are warm-worked to produce high strength levels. The warm working also produces significant sensitization. This drill collar was used for deep ocean geologic core sampling, with seawater as the drilling fluid. The sensitized microstructure developed IGSCC during the high-stress drilling operations even though temperatures never exceeded surface ambient. The drill collar eventually parted, resulting in the loss of the bottom-hole assembly.
Remedy: Use low-carbon (0.03% max.) alloys, and impose residual compression stresses on the inside diameter with mill finishing practice.

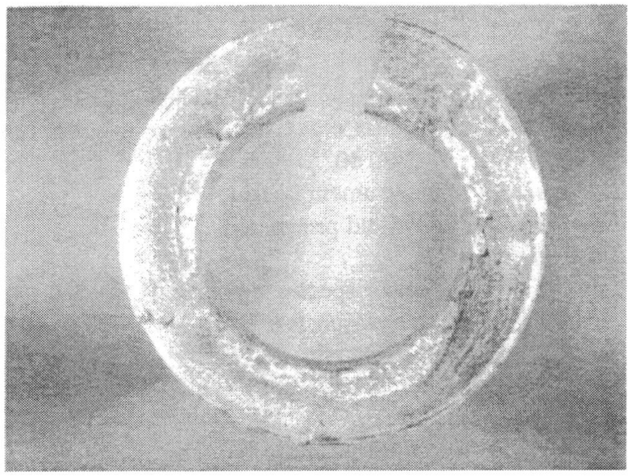

Figure 4.48(a) - Photograph

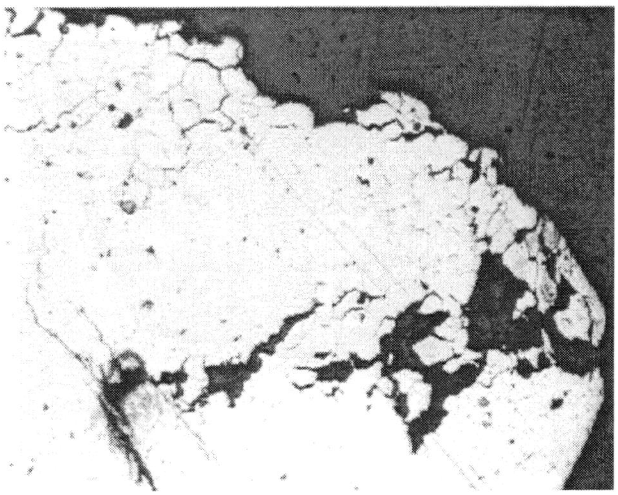

Figure 4.48(b) - Micro

CASE HISTORY 4.49

Phenomenon: corrosion fatigue
Group/ Type: AISI 304 SS (UNS S30400)
Equipment: steam sparger
Environment: chloride-bearing process water
Conditions:
(a) major species: water and steam
(b) minor species: vinyl chloride
(c) temperature: 200 psi steam temperature
(d) pressure: 200 psi
(e) aeration: none
Time to failure: 9 months
Illustration: macro
Comments: This 2-in. diameter AISI 304 SS pipe with 0.5-in. thick wall supported a 30 lb CF8M cast SS sparger head. After nine months intermittent service, the sparger pipe failed at its base in the weld fillet. The figure shows the fracture with the relatively flat, beachmarked surface characteristic of fatigue. The source of the cyclic stress was believed to be vibration and the weight of the sparger head.
Remedy: Use a smaller sparger head and switch to 17-4pH SS hollow bar (condition H1150) for higher tensile strength.

Figure 4.49 - Macro

CASE HISTORY 4.50

Phenomenon: sulfide stress cracking
Group/ Type: SS, austenitic 21Cr, 13Ni, 2.3Mo, 0.4% Cu with nitrogen, cold-worked to HRC 40
Equipment: instrument housing
Environment: oilfield production
Conditions:

(a) major species: methane, H_2O
(b) minor species: 9% CO_2, 16.3% H_2S, 67,000 ppm Cl-
(c) temperature: 221°F (105°C)
(d) pressure: 2,315 psig
(e) aeration: anaerobic

Time to failure: 7.5 days
Illustration: macro
Comment: Austenitic SS can become extremely sensitive to mixed chloride/sulfide stress-cracking if it is cold-worked. This instrument housing suffered rapid transgranular environmental cracking in a sour gas well.
Remedy: Use nickel alloys or nickel-copper alloy housings.

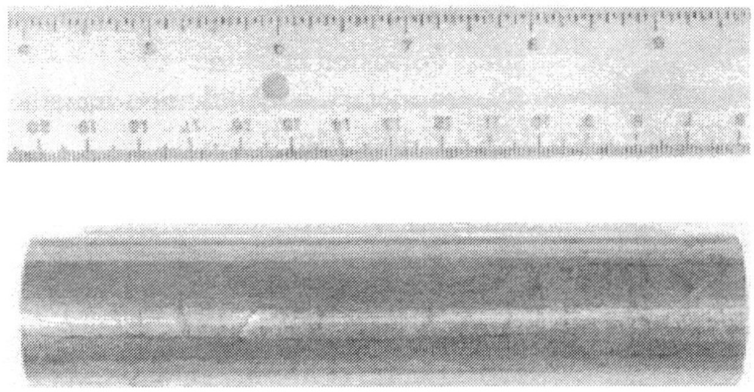

CASE HISTORY 4.51

Phenomenon: chloride/sulfide SCC
Group/ Type: stainless steel AISI 316 (UNS S31600)
Equipment: sulfur recovery pit
Environment: molten sulfur fumes
Conditions:

(a) major species: sulfur
(b) minor species: water, chlorides
(c) temperature: 300°F (149°C)
(d) pressure: atmospheric
(e) aeration: not determined

Time to failure:
Illustration: macro, SEM fractograph
Comment: Occasional fires in a molten sulfur pit were quenched with live steam. The mixture of steam and sulfulr vapors created a slightly acidic environment (pH 5 beneath deposited scales). After several months of such services, a AISI 316 SS I-beam was found cracked through its lower flange and up into the web for distance of ~ 8 in. Fractography and metallography showed that the cracking was predominantly transgranular and highly branched, indicative of SCC. Although no more than 25 ppm were detected in any of the process inlet streams, 8,000 ppm chloride were detected on the exposed fracture surface shown in the figure. The tolerance of SS for low levels of chlorides drops drastically with lowered pH and the presence of sulfur compounds.
Remedy: The beam was replaced with a concrete I-beam coated with an acid-resistant mortar.

Figure 4.51 (a) Macro and (b) SEM fractograph.

CASE HISTORY 4.52

Phenomenon: SCC
Group/ Type: AISI 304 SS (UNS S30400)
Equipment: check valve
Environment: water treating chemical injection
Conditions:

(a) major species: 7% NaOH in water
(b) minor species: 4% chelant
(c) temperature: 550°F (288°C)
(d) pressure: 20 psi
(e) aeration: not determined

Time to failure: 20 days
Illustration: macro
Comment: For convenience, this water-treatment chemical injection line was heat-traced with steam from a nearby superheated 200 psi steam line. The superheated steam was well in excess of threshold temperatures for caustic SCC of stainless steel. Complete rupture of the valve body occurred at longitudinal SCC cracks.
Remedy: Valve was replaced in kind and steam-traced with 70 psi [270°F (132°C)] steam. No further failures occurred in 18 months of service.

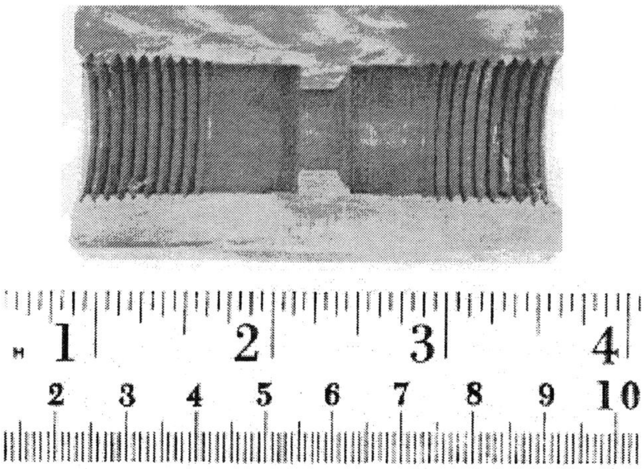

Figure 4.52(a) - Valve body cross section. Fracture surface is at top.

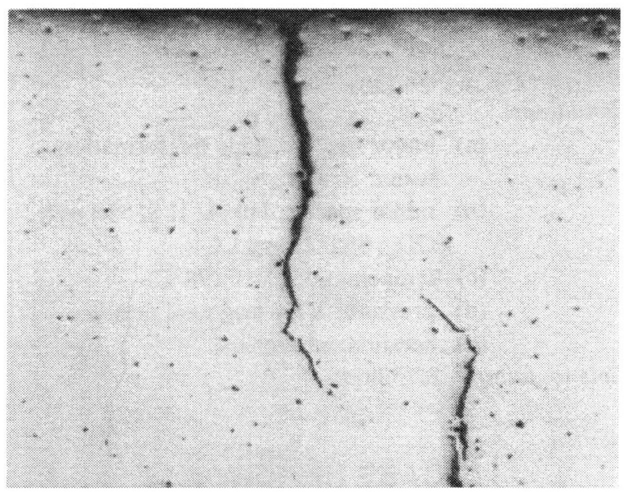

Figure 4.52(b) - Unetched metallographic section showing secondary cracks, 150X.

CASE HISTORY 4.53

Phenomenon: SCC
Group/ Type: SS duplex austenitic-ferritic 25Cr
Equipment: laboratory corrosion coupon
Environment: simulated natural gas well
Conditions:

(a) major species – liquid: 19% NaCl brine; gas – 3% H_2S, balance CO_2
(b) minor species:
(c) temperature: ambient
(d) pressure: 250 psig
(e) aeration: anaerobic

Time to failure: 30 days
Illustration: none
Comment: C-rings and DCB specimens fully exposed in this simulated sour gas environment showed no crack growth. However, numerous stress corrosion cracks initiated under the TFE-fluorocarbon crevice on a weight-loss coupon. Cracking occurred because of the combination of the low pH in the crevice and the thin layer of cold-worked material left from machining the samples from their original tubes.
Remedy: Use austenitic high-nickel alloys for extreme sour service.

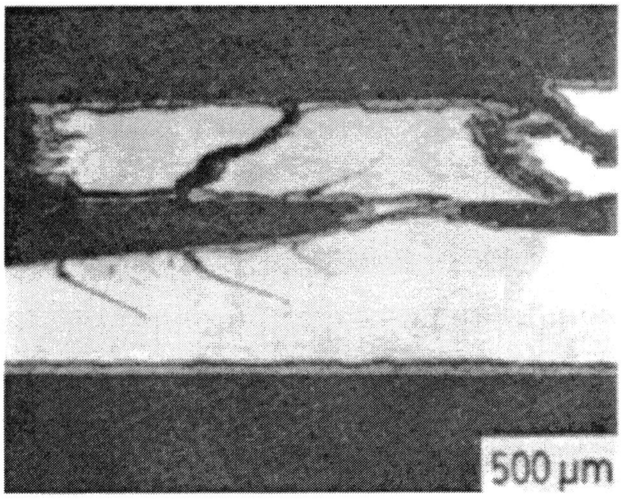

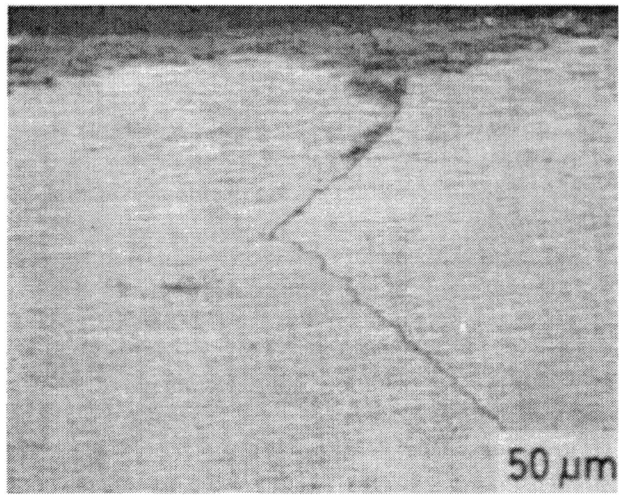

CASE HISTORY 4.54

Phenomenon: sulfide/chloride stress cracking
Group/Type: duplex SS, 23 Cr, 7Ni, 2Mo, 2Cu
Equipment: wireline for gas well pressure recorder
Environment: sour gas well
Conditions:
(a) major species: crude oil, formation water
(b) minor species: 110 psi H_2S, 550 psia CO_2, 85,000 ppm Cl-
(c) temperature: 194°F (90°C)
(d) pressure: 2,750 psig total pressure
(e) aeration: anaerobic
Time to failure: 8.75 hours
Illustration: macro, photomicrograph
Comment: This 0.092-in. diameter wireline was stressed to 360 MPa (52 ksi) by hanging weight. After a little more than eight hours, the wireline parted just above the liquid level in this natural gas well. The wire had been cold-worked to HRC 46 for mechanical strength. Duplex SSs showed a minimum in environmental cracking at ~ 200°F (93°C). At that temperature, cracks can propagate by the synergistic effects of both hydrogen-assisted cracking and anodic SCC.
Remedy: Use high chromium, high molybdenum nickel-based alloy for wireline in sour oil and gas wells.
Reference: S.W. Ciaraldi, "Materials Failures in Sour Gas Service," CORROSION/85, paper no. 217 (Houston, TX: NACE, 1985).

CASE HISTORY 4.55

Phenomenon: SCC
Group/ Type: AISI 304 SS (UNS S30400)
Equipment: thermowell
Environment: crude oil
Conditions:
(a) major species: crude oil
(b) minor species: H_2O, NaCl, trace H_2S
(c) temperature: ambient to 95°F (35°C)
(d) pressure: 200 psi
(e) aeration: none
Time to failure: 9 months
Illustration: SEM fractograph
Comment: This cold-worked SS tube was brazed into a threaded pipe connection and used as a thermowell. Failure occurred after nine months in a sour crude oil pipeline. SEM examination showed the quasi-cleavage facets apparent on SCC failures; 300 series SS often are used in

the solution-annealed condition in sour gas environments. However, upon cold-working, some reversion to martensite occurs, rendering the materials susceptible to environmental cracking in H_2S.
Remedy: Use solution-annealed SS according to NACE Standard MR0175-96, "Sulfide Stress Cracking Resistant Metallic Materials for Oilfield Equipment," for sour oil and gas service.

CASE HISTORY 4.56

Phenomena: SCC, pitting
Group/ Type: AISI 316 SS
Equipment: cooling coils in molten sulfur pit
Environment: molten sulfur
Conditions:
- a) major species: molten s sulfur
- b) minor species: well water or condensate
- c) temperature: 270ºF (132°C)
- d) pressure: 1 atmosphere
- e) aeration:

Time to failure: 10 years
Illustration: macro, closeup macro, micro
Comment: Molten sulfur free of water and chlorides can readily be handled in austenitic stainless steels. However, in this case the concrete cover over a below-grade molten sulfur pit was occasionally washed down with well water containing about 1,000 ppm chloride ion. Some of the wash water ran down gaps between the cooling coils and the concrete cover. When the water contacted hot sulfur fumes the two combined to create chloride-contaminated sulfurous acids, causing severe pitting and stress-corrosion cracking on the coils above the sulfur liquid level.
Remedy: Eliminate water from the pit with a coated carbon steel sleeve around the cooling coil opening.

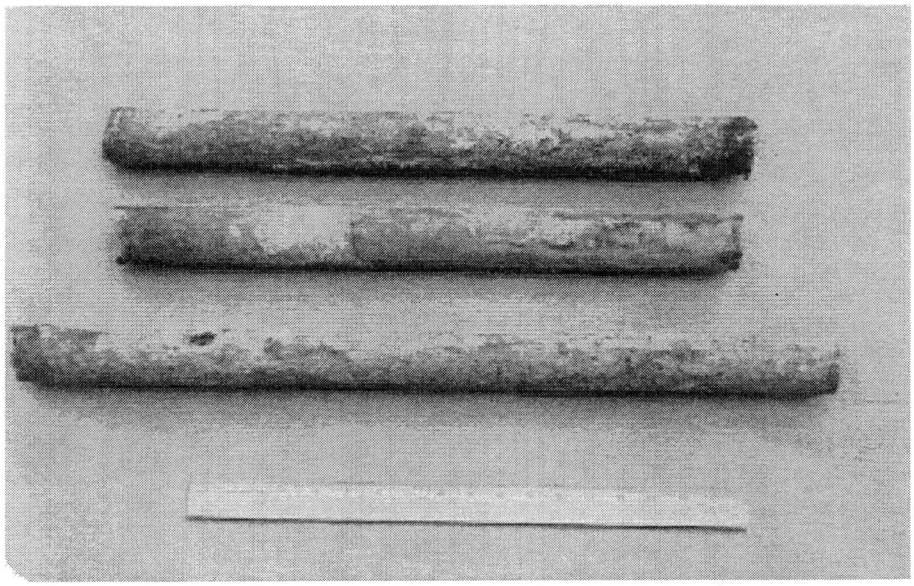

Figure 4.56 (a) - Macro

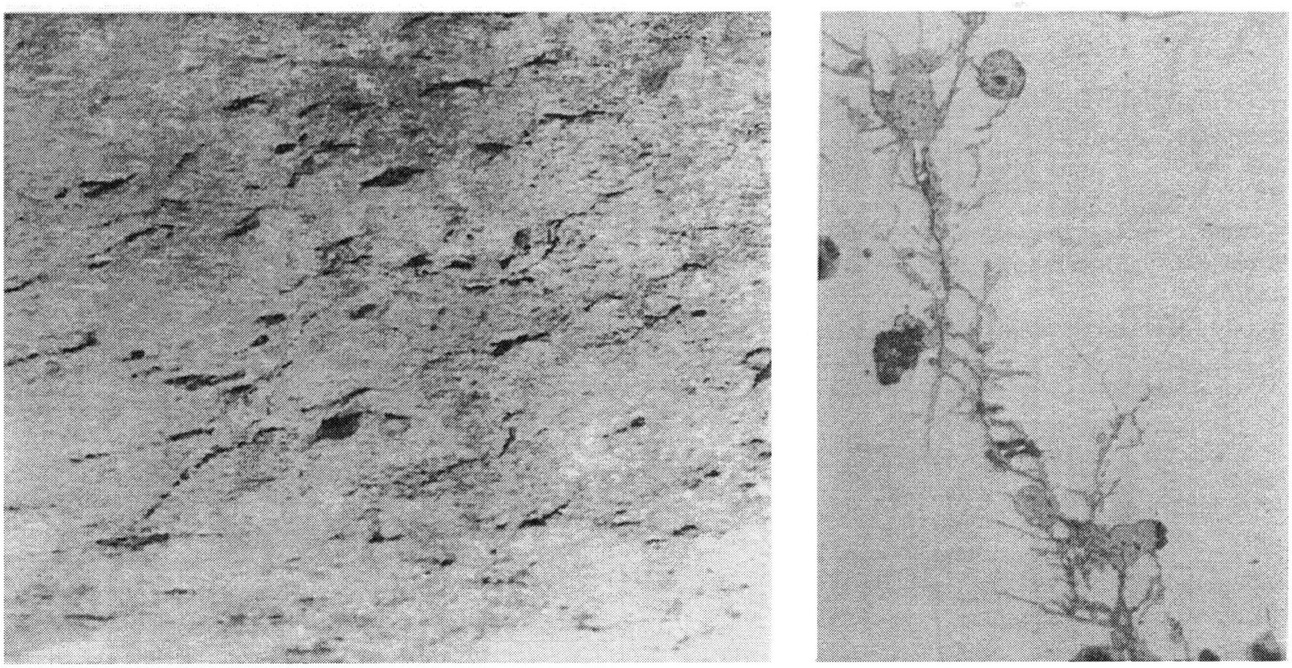

Figure 4.56- (b) Macro closeup of outer surface and (c) photomicrograph

CASE HISTORY 4.57

Phenomenon: stress-corrosion cracking
Group/Type: AISI type 316 SS
Equipment: thermowell
Environment: sour crude oil
Conditions:
a) major species: crude oil
b) minor species: 2% sulfur, 0.2% water with 15,000 ppm NaCl
c) temperature: Arabian Gulf ambient (40-130ºF [4.4-54°C])
d) pressure: 550 psig
e) aeration: anaerobic

Time to failure: 12 years
Illustration: photomicrograph
Comment: The lower threshold temperature for stress-corrosion cracking of austenitic stainless steels can be depressed below the widely used 60ºC (140ºF) figure by a number of factors, including cold-work and the presence of H_2S in the environment. Cold-worked AISI 316 stainless steel was chosen for this thermowell to overcome a fatigue cracking problem. However, the thermowell suffered stress-corrosion cracking in the ambient temperature, wet sour crude oil.
Remedy: Use duplex stainless steel thermowells, which have good SCC resistance and good fatigue resistance in the solution-annealed condition.

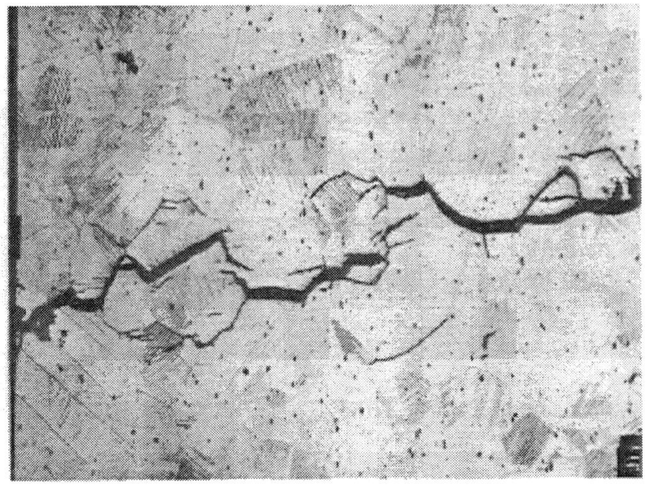

Figure 4.57 - Photomicrograph, low-temperature SCC of cold-worked stainless steel thermowell.

CASE HISTORY 4.58

Phenomenon: SCC
Group/ Type: copper alloys/ DHP copper, CA 122 (UNS C12200)
Equipment: heat exchanger tubes
Environment: shop fabrication
Conditions:
(a) major species: shop air, steam during cleaning
(b) minor species: water vapor, oxygen, amine inhibitors
(c) temperature: ambient [250°F (121°C)] during steam cleaning
(d) pressure: atmospheric
(e) aeration: saturated

Time to failure: 1 to 6 weeks
Illustrations: SEM fractograph
Comment: Twenty percent cold-drawn copper tubes were brazed into brass tubesheets in a fabrication shop, then acid-rinsed and steam-cleaned to remove brazing fluxes and high-temperature oxides. After shop storage for one to six weeks, 3-18% of the tubes were found to be cracked and leaking. All leaks were in the middle of the tubes rather than near the brazed ends. Investigation revealed that the proprietary amine inhibitor in the steam system broke down at high temperatures to liberate small amounts of ammonia. The DHP copper tubes were not prone to cracking near the tubesheets, where the heat of brazing relieved the stresses of cold-work, but severe ammoniacal SCC resulted in the center of the tubes where residual stresses of cold-work were still present.
Remedy: The inhibitor was switched to a borax-sodium nitrite mixture, and no further incidents of cracking were observed in more than four years.

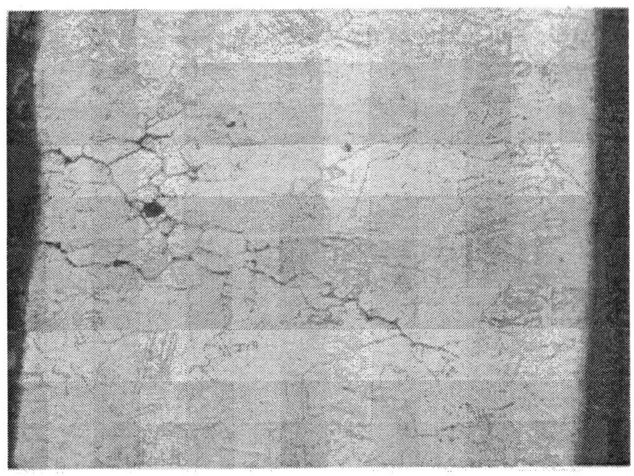

Figure 4.58 - NH_4OH etch, 250X.

CASE HISTORY 4.59

Phenomenon: SCC
Group/ Type: copper alloys; commercial tin bronze
Equipment: air filter cartridge closure bolt
Environment: industrial atmosphere
Conditions:
(a) major species: air, water vapor
(b) minor species: ammonia vapor
(c) temperature: ambient
(d) pressure: atmospheric
(e) aeration: saturated

Time to failure: 10 years
Illustration: macro, micro
Comment: This air filter cartridge hold-down bolt was installed on a 100 psi compressed air line. After 10 years in service, the bolt broke, allowing the air pressure to propel the cartridge across a control room, narrowly missing a nearby operator. Investigation showed an intergranular crack path indicative of SCC. Ammonia vapor releases in a refrigerant unit 100 ft upwind were suspected as the cause.
Remedy: Replace with AISI 304 SS (UNS S30400).

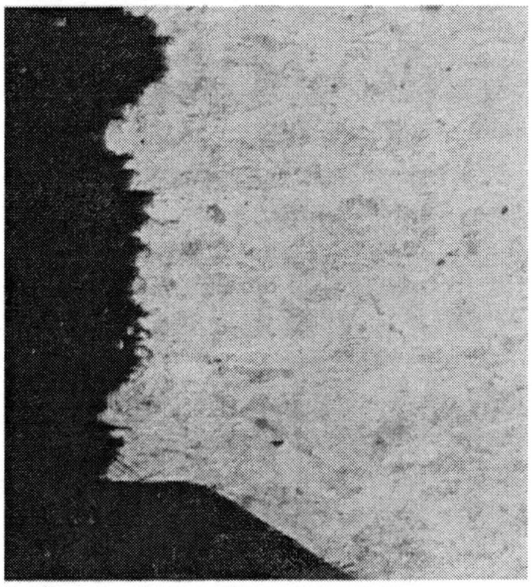

Figure 4.59 - (a) Macro and (b) micro.

CASE HISTORY 4.60

Phenomenon: corrosion fatigue
Group/Type: copper-based Alloys, Ni-Al-bronze
Equipment: ship's propeller
Environment: seawater
Conditions:
(a) major species: water, 3.5% NaCl
(b) minor species: carbonates, bicarbonates, fluorides, etc.
(c) temperature: ambient
(d) pressure: ambient (water depth 20 ft)
(e) aeration: saturated

Time to failure: 9 months at 112 rpm
Illustration: macro, fracture surface
Comment: A freighter lost a propeller tip in mid-voyage. When examined, the propeller's fracture surface showed chevron marks pointing back to a smooth, beachmarked zone of corrosion fatigue. Beachmarks and shear steps indicated a series of fracture initiation sites along the aft side of the propeller blade. All of the initiation sites occurred in an area that had been welded to repair extensive dealloying. After initiation on the aft (tension) surface, the fatigue crack grew through a dealloyed area, which escaped repair to reach critical size.
Remedy: Inspect with ultrasound to ensure a sound repair, shot-peen the tension surface, or switch to higher strength precipitation-hardening SS propeller.

Figure 4.60 - Macro, fracture surface

CASE HISTORY 4.61

Phenomenon: SCC
Group/ Type: copper-based, 90-10 cupro-nickel (UNS C70600)
Equipment: moisture separator reheater tubing
Environment: steam
Conditions:
(a) major species: steam
(b) minor species: hydrazine, cyclohexyl lamine and sodium phosphate water treatment chemicals
(c) temperature: 392°F (200°C)
(d) pressure: not given
(e) aeration: none

Time to failure: 3 years
Illustration: micro, direct imaging mass analyzer (DIMA) spectrograph
Comment: Several tube leaks were attributed to IGSCC after some months of a water treatment, including combined hydrazine, cyclohexylamine, and sodium phosphate. DIMA examination revealed evidence of nitrogen ions and the ammonia radical in the corrosion deposits at the crack tip. Consequently, the failure was attributed to ammonia SCC.
Remedy: Not specified
Reference: F.W. Pement, I.L.W. Wilson, S.D. Reynolds Jr., W.D. Fletcher, *Materials Performance*, 16, 1 (1977): pp. 26-39.

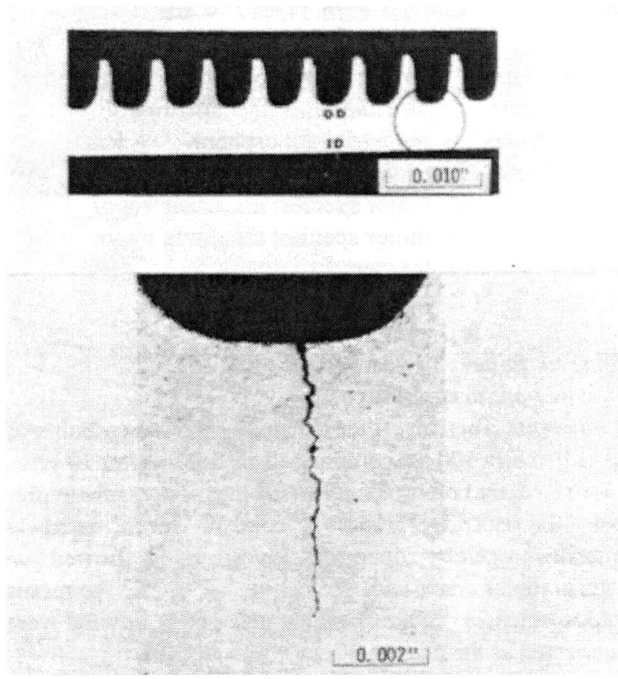

Figure 4.61 - Micro, direct imaging mass analyzer (DIMA) spectograph.

CASE HISTORY 4.62

Phenomenon: SCC
Group/ Type: copper alloys, tin bronze
Equipment: axe
Environment: wet soil
Conditions:
(a) major species: earth, humus
(b) minor species: water, organic decomposition products
(c) temperature: ambient
(d) pressure: 1 atm
(e) aeration: restricted but not totally anaerobic

Time to failure: unknown (total life 4,000+ years)
Illustration: micro
Comment: The formation of corrosion products causes a substantial volume change in metals, and if the item is under constraint, the buildup of such corrosion products can cause significant stresses. This bronze axe was recovered from a Babylonian tomb, where it had lain under no apparent applied stresses for ~ 4,000 years. Metallography revealed several stress corrosion cracks (See Figure) initiating beneath corrosion products wedged into locations where they could not expand and flake off. This specimen is believed to be the oldest known example of SCC found in any alloy.

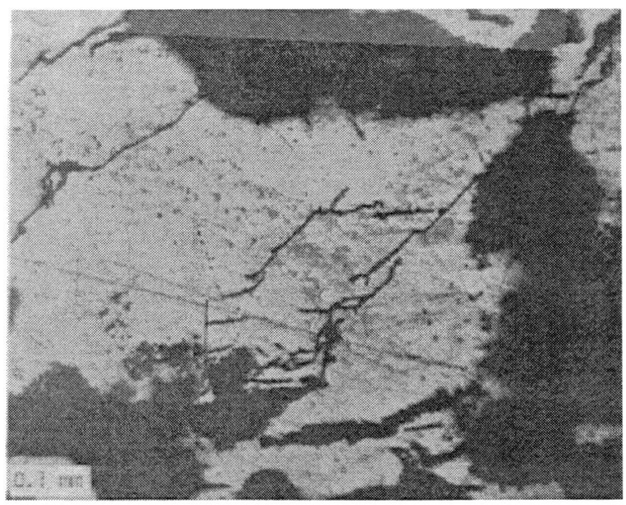

Remedy: Not applicable
Contributor: G. Ugiansky, NBS

CASE HISTORY 4.63

Phenomenon: hydrogen sulfide stress-cracking
Group/ Type: nickel-based alloy, UNS N06600
Equipment: compressor valve springs
Environment: refinery gas
Conditions:
(a) major species: H_2, CH_4, CO_2, CO
(b) minor species: H_2S
(c) temperature: 125°F (52°C)
(d) pressure: not given
(e) aeration: anaerobic
Time to failure: not given
Illustration: SEM fractographs
Comment: Nickel-based alloys have high resistance to SSC unless hardened above HRC 35 and coupled to steel. This cold-worked alloy 600 (UNS N06600) valve spring failed quickly in a compressor handling sour refinery gas. Brittle quasi-cleavage features were observed on SEM examination.
Remedy: A switch to cobalt-based alloy MP35N springs eliminated sulfide stress cracking of springs.

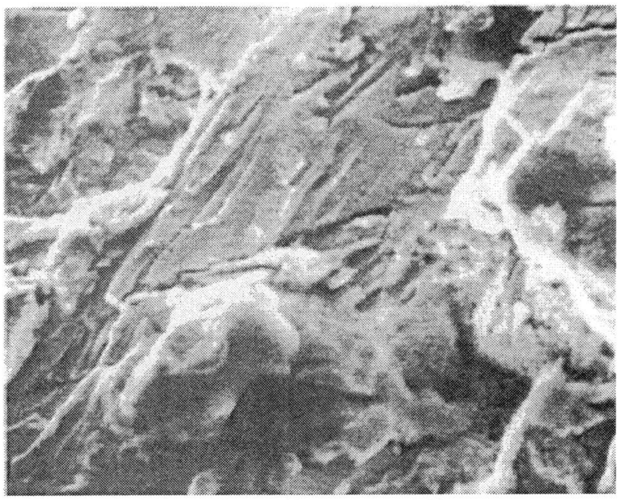

Figure 4.63 - SEM fractograph, 1,000X.

CASE HISTORY 4.64

Phenomenon: SCC
Group/ Type: nickel-based alloy UNS N10276
Equipment: autoclave
Environment: simulated oil-well acidizing solution
Conditions:
(a) major species: 13% HCl, 2% HF
(b) minor species: -
(c) temperature: 325°F (162.5°C)
(d) pressure: 80 psig
(e) aeration: anaerobic
Time to failure: 66 h
Illustration: photomacrograph
Comment: UNS alloy N10276 generally is considered immune to SCC across a wide range of environmental conditions. However, in this case, SCC occurred on this laboratory autoclave head after exposure to a simulated hydrocarbon production acidizing solution consisting of a mixture of HCl and HF. Cracking was confined to the vicinity of a seal weld for a gas inlet port and was, therefore, assumed to be associated with residual stresses from welding.
Remedy: Shot peen or use TFE-fluorocarbon autoclave for HCl-HF service.

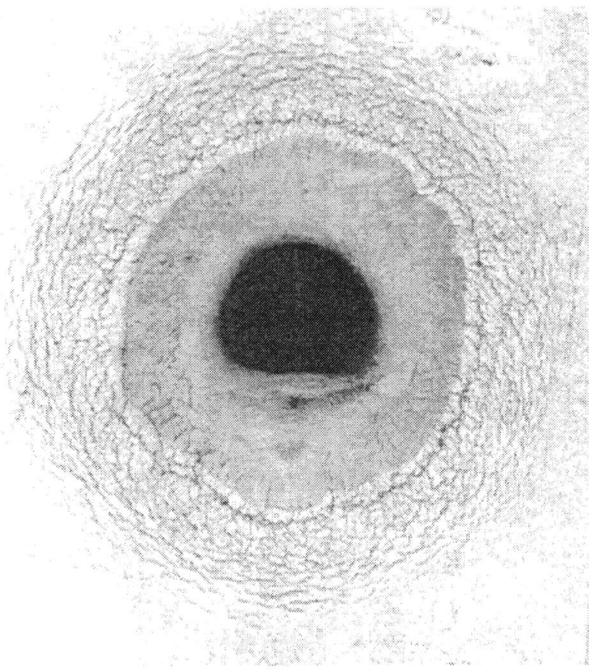

Figure 4.64 - Macro

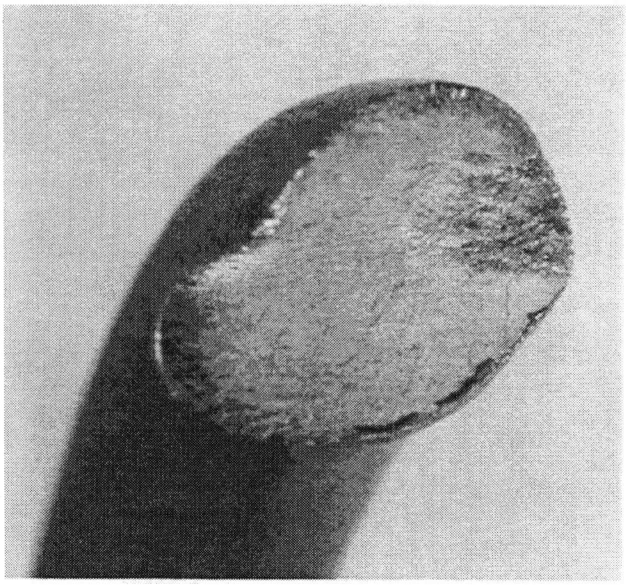

CASE HISTORY 4.65

Phenomenon: SCC
Group/ Type: titanium alloy/13Cr-11V-3Al (UNS R58010)
Equipment: aircraft slat actuator spring
Environment: atmosphere + exhaust gas
Conditions:
(a) major species: air, water vapor, jet engine exhaust
(b) minor species: chlorides, SO_2, NOx
(c) temperature: not determined
(d) pressure: atmospheric
(e) aeration: saturated

Time to failure: ~ 6 months
Illustration: macro, SEM fractograph
Comment: Titanium alloys vary widely in their SCC resistance depending on alloying content, toughness and strength. This spring material failed prematurely on a commercial jetliner in a location where it would be exposed alternatively to rainwater and jet engine exhaust gases.
Remedy: Shield the spring from water and jet engine exhaust.

CASE HISTORY 4.66

Phenomenon: anodic SCC
Group/ Type: titanium alloys, grade 2 commercially pure (UNS R50400)
Equipment: shell-and-tube heat exchanger
Environment: anhydrous ethanol + HCl
Conditions:
(a) major species: 62% ethanol, 14% NaCl, 24% other organics
(b) minor species: 128 ppm HCl (nominal)
(c) temperature: 50°F (10°C)
(d) pressure: not given
(e) aeration: not determined
Time to failure: 3 weeks
Illustration: schematic
Comment: Intergranular stress corrosion cracks grew from the process side of this heat exchanger tubesheet. Laboratory slow strain rate and U-bend tests confirmed that anhydrous ethanol can cause SCC on titanium if the ethanol contains more than ~ 0.4% HCl. The presence of NaCl apparently aggravates the attack.
Remedy: Replace with a graphite heat exchanger
Reference: E.L. Liening, *Materials Performance*, 22, 11 (1983): pp 37-44.

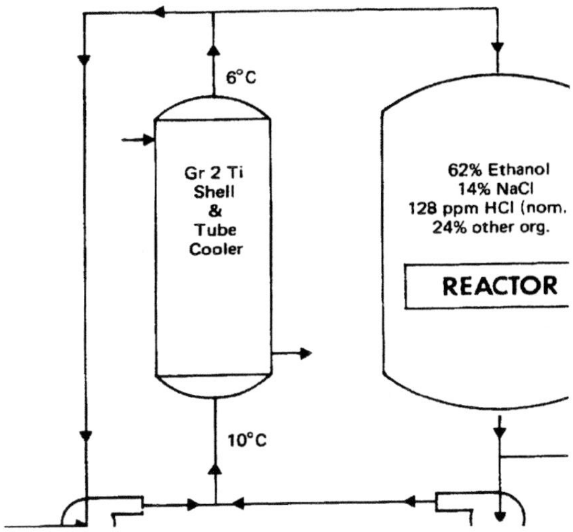

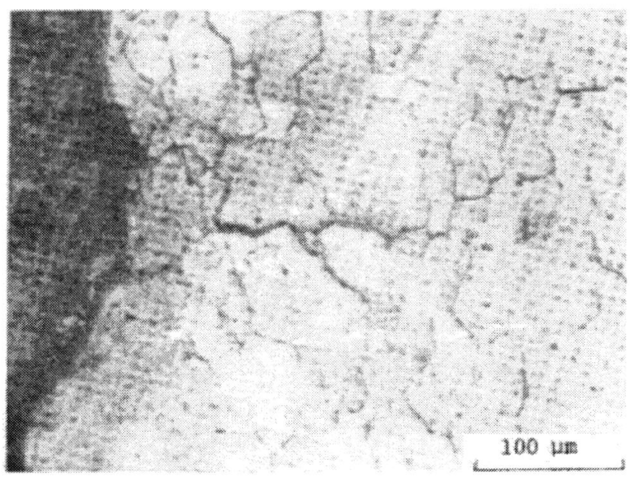

CHAPTER 5

Erosion-Corrosion, Cavitation, and Fretting

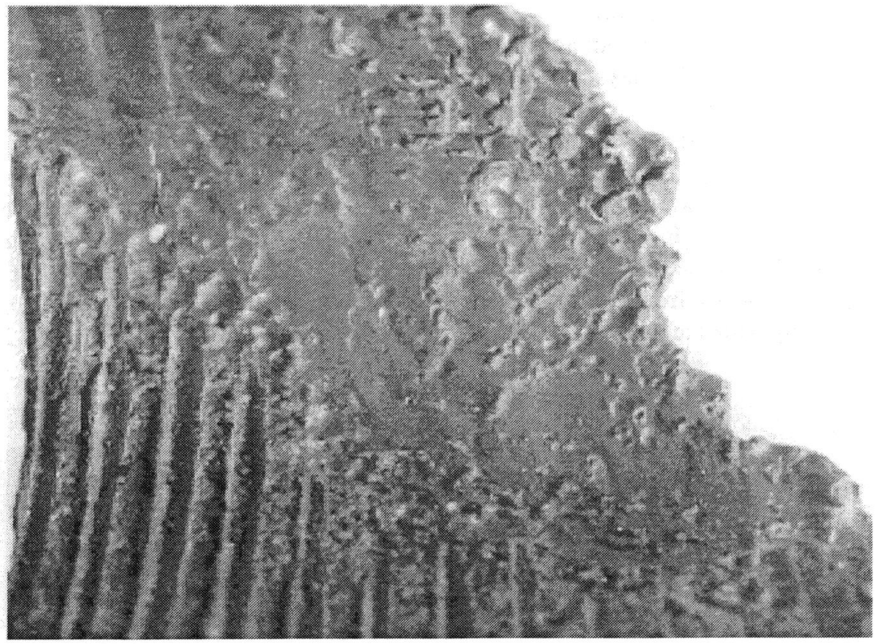

CASE HISTORY 5.1

Phenomenon: erosion-corrosion
Group/ Type: ferrous: carbon steel
Equipment: heat exchanger tubes
Environment: shell-side steam, tube-side mineral slurry
Conditions:
- (a) major species: freshwater
- (b) minor species: inert mineral particles
- (c) temperature: 104°F (40°C) inlet, 194°F (90°C) outlet
- (d) pressure: not given
- (e) aeration: saturated

Time to failure: 10 years
Illustration: macro
Comments: This photograph shows a classic case of erosion-corrosion on the inlet end of a heat exchanger. The slurry of water and fine mineral particles did not attack the heat exchanger heads. However, for the first foot of length on the inlet tube ends, there was severe metal loss, leading to penetration. The corroded surface showed the classic horseshoe-shaped flow pattern characteristic of erosion. After the first foot or so of tube length, no significant metal loss occurred in the tubes or on the tubes' outlet ends.
Remedy: Use TFE-fluorocarbon inserts on the inlet ends, or reverse the heat exchanger bundle to double the life.

CASE HISTORY 5.2

Phenomenon: erosion-corrosion
Group/ Type: ferrous, API N-80 tubing grade.
Equipment: oil well tubing
Environment: crude oil
Conditions:
(a) major species: 13,000 bbl/day crude oil, 20.5 mmcf/day CH_4
(b) minor species: 90 psi CO_2, 1% H_2O in emulsion
(c) temperature: 200°F (93°C)
(d) pressure: 900 psi wellhead flowing pressure
(e) aeration: anaerobic
Time to failure: 241 days
Illustration: macro
Comment: Corrosion is not normally expected in oil wells when no water is produced. In this well, although there was no separate water phase, ~ 1% water was emulsified in the oil. The high CO_2 partial pressures produced a pH measured as 6.0 at the separator. Velocities in this tubing string ranged from 21 to 26 ft/s. During a workover, the string parted at a depth of 1,400 ft in an area where the tubing had been completely perforated by corrosion occurring at rates of ~ 400 mpy.
Remedy: Inject corrosion inhibitor on a continuous basis.
Reference: C.J. Houghton, R.V. Westermark, *Materials Performance*, 22, 1 (1983): pp. 16-22.

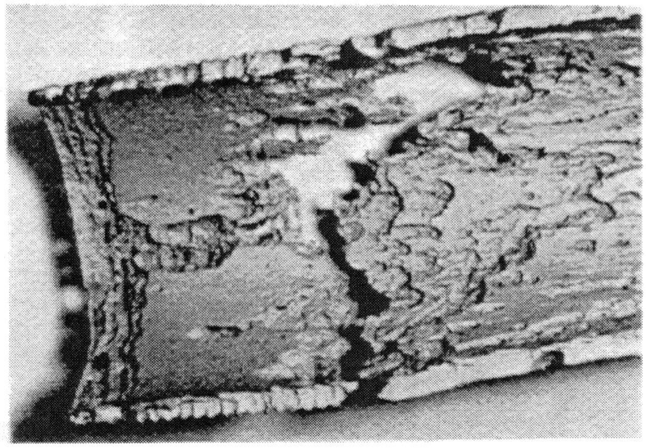

Figure 5.2 - 4½ in. L-80 tubing, depth 5,700 ft.

CASE HISTORY 5.3

Phenomenon: impingement
Group/ Type: ferrous, AISI 4140 (UNS G41400)
Equipment: flow coupling
Environment: natural gas condensate production
Conditions:
(a) major species: CH_4 - 37.5 mmcf/day, oil 2,400 - bbl/day
(b) minor species: water - 25 bbl/day, CO_2 - 60 psi
(c) temperature: 175°F (79°C) flowing
(d) pressure: 1,117 psig wellhead flowing pressure
(e) aeration: anaerobic
Time to failure: 35 months
Illustration: macro
Comment: The fluid produced from this well was, by itself, only moderately corrosive. Surface monitoring of weight loss coupons showed average corrosion rates of 17 mpy (430/y). This flow coupling parted immediately above the subsurface safety valve. The flowing velocity range for the tubing was 40 to 125 ft/s (12.2 to 38.1 m/s). However, as the fluid exited the 2-in. inner diameter safety valve into the 3.9-in. tubing, the exit velocity was over 300 ft/s (91 m/s). Severe impingement attack was the result.
Remedy: Flow coupling was replaced with 13Cr martensitic SS.
Reference: C.J. Houghton, R.V. Westermark, *Materials Performance*, 22, 1 (1983): pp. 16-22.

Figure 5.3 - 4½ in. flow coupling and subsurface safety valve.

CASE HISTORY 5.4

Phenomenon: erosion-corrosion
Group/ Type: carbon steel
Equipment: transfer pipe line
Environment: sulfuric acid
Conditions:
(a) major species: 98% sulfuric acid
(b) minor species:
(c) temperature: ambient
(d) pressure: not specified
(e) aeration: saturated
Time to failure: less than one week
Illustration: none
Comment: Concentrated sulfuric acid was transferred into a batch process through a steel pipe line. To reduce the transfer time from one hour to the desired time of 15 minutes, a stronger pump was installed. The new pump increased the velocity from ~ 2 ft/s to more than 5 ft/s. The velocity increase stripped off protective sulfate films that prevented accelerated attack on the pipe line, and rapid perforation from corrosion resulted.
Remedy: Maintain velocity below 5 ft/s in concentrated sulfuric acid service.
Reference: O.W. Siebert, "Classic Blunders in Corrosion Protection Revisited," CORROSION/83, paper no. 24 (Houston, TX: NACE, 1983)

CASE HISTORY 5.5

Phenomenon: cavitation
Group/ Type: ferrous/carbon steel
Equipment: heat exchanger
Environment: shell-side steam, tube-side hydrocarbons + HF
Conditions:
(a) major species: steam
(b) minor species: not determined
(c) temperature: 330.8 to 356°F (166 to 180°C)
(d) pressure: 90 to 130 psi
(e) aeration: not determined
Time to failure: first failure, three months; second failure, one week.
Illustration: macro
Comment: Leaks in a hydrocarbon-HF heat exchanger led to rapid penetration of the shell by general corrosion from the HF. However, investigation showed that the initial leaks were caused by pitting of the 3/4-in. 12 gauge carbon steel tubes from the steam side. The pits occurred only on the top tube row, on the side opposite the steam inlet. Up to 0.016 in. of wear was observed at the baffles, and all the pits occurred beneath the two center baffles. The shape and location of the pits suggested this unusual attack was related to vibration and wear, setting up cavitation conditions, where bubbles could form and collapse beneath the baffles on the worn tube surface.

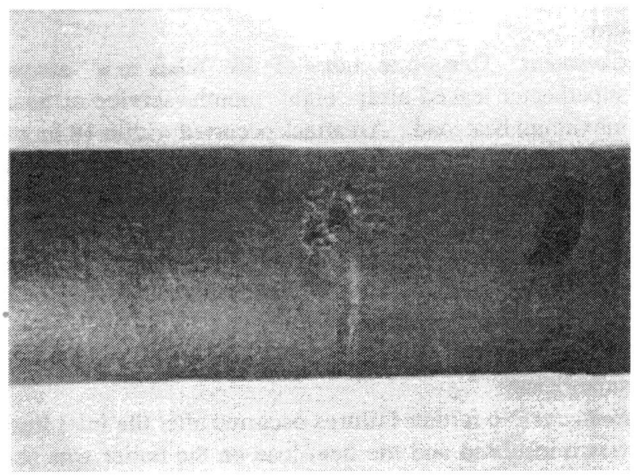

Remedy: Eliminate tube vibration by balancing baffle spacing and flow rates.

CASE HISTORY 5.6

Phenomenon: erosion-corrosion
Group/ Type: ferrous/ SA178 grade A carbon steel
Equipment: finned superheater tubes
Environment: tubeside steam, fin-side natural gas exhaust
Conditions:
(a) major species: steam
(b) minor species: water droplets, 400 to 800 ppm $CaCO_3$, 10 to 60 ppm PO_4, 20 to 100 ppm Cl-, 20 to 80 ppm sulfite, 1 to 4 ppm Fe
(c) temperature: 530°F (276°C)
(d) pressure: 265 psi
(e) aeration: not determined
Time to failure: 8 months
Illustration: macro of tube cross-section and SEM micro.
Comment: Thirty-one out of 36 tubes in a steam superheater leaked after eight months service at near maximum heat load. All attack occurred within 18 in. of the tubes' inlet ends. The inner surfaces of the tubes showed evidence of erosion-corrosion (Figures, macro) and a loose, flaky black magnetite scale (Figures, SEM micro) covered much of the affected surfaces. Failure was attributed to the failure of the insulation on the inlet line, which allowed condensation and droplet formation in the inlet line between the steam drum and the superheater.
Remedy: No further failures occurred after the inlet line was reinsulated and the heat load on the boiler was reduced.

CASE HISTORY 5.7

Phenomenon: erosion-corrosion
Group/Type: carbon steel, ASTM A106 Gr B, 4 inch Sch. 80 pipe
Equipment: air-cooled condenser nozzles
Environment: hydrocracker reactor effluent
Conditions:
a) major species: gasoline, H_2S
b) minor species: 0-5% wash water, 4-9% NH_4HS in separator
c) temperature: 24 to 50ºC (-4.4 to 10°C)
d) pressure: 2,500 psig
e) aeration: 1-2 ppm O_2 in the wash water
Time to failure: 3 years
Illustration: sketch of nozzle

Comment: The outlet nozzles on these air-cooled condensers were corroding at rates up to 170 mils per year due to the formation of ammonium bisulfide in the reactor effluent. The ammonium-bisulfide is corrosive in water and erosive if it drops out of solution as particulates. Eliminating the corrosion is a complex balancing act requiring the control of critical process variables between certain limiting ranges.
Remedy: Weld overlay the nozzles with alloy 625 (UNS N06625). To eliminate associated corrosion on the condenser tubes and piping, reduce the oxygen content in the wash water to 20 ppb, increase the wash water injection rate to give <5% NH_4HS in the separator, and increase the condenser outlet temperature to 50ºC to minimize the for mation of solid NH_4HS.

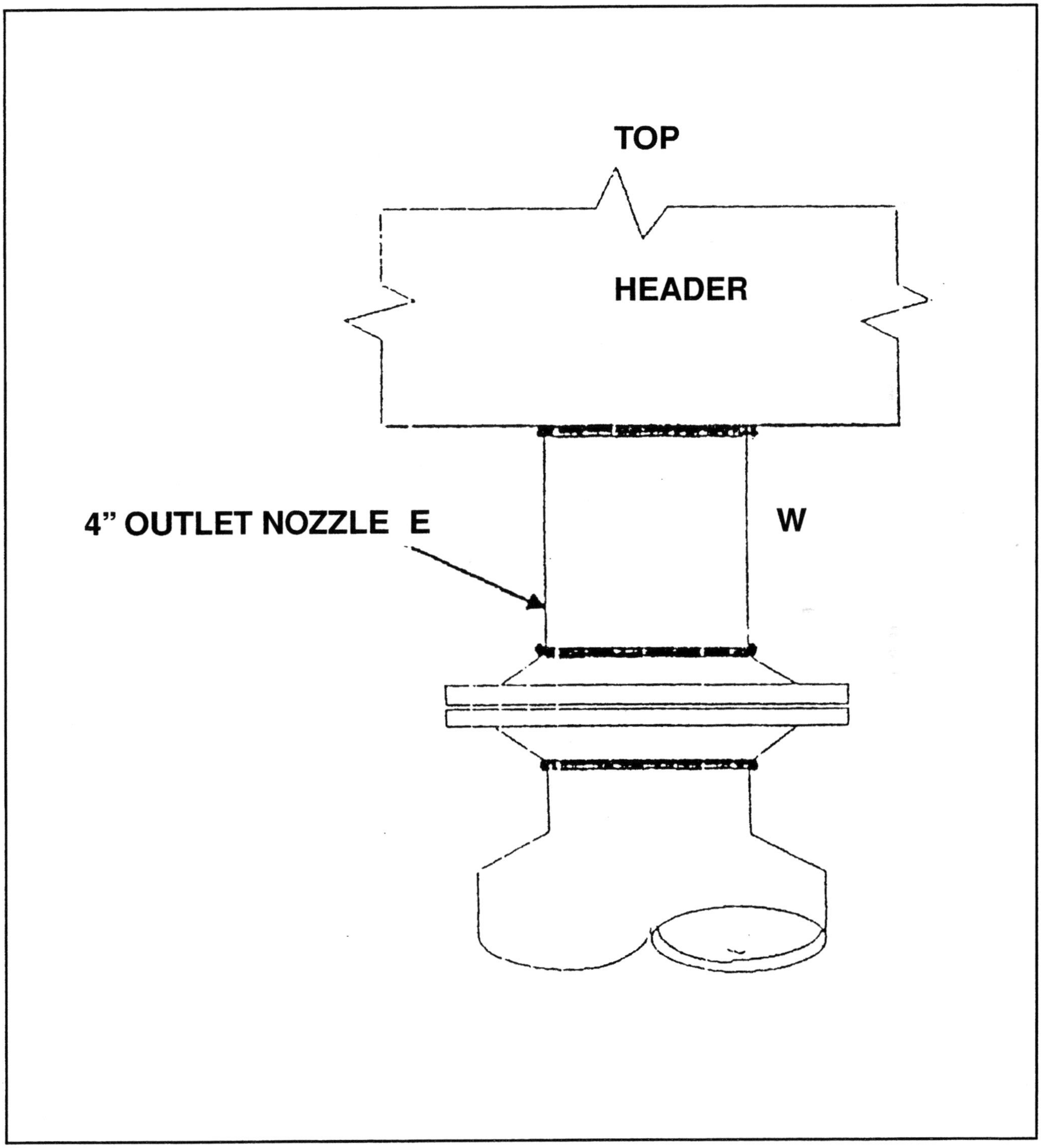

Figure 5.7 - Sketch of eroded nozzle.

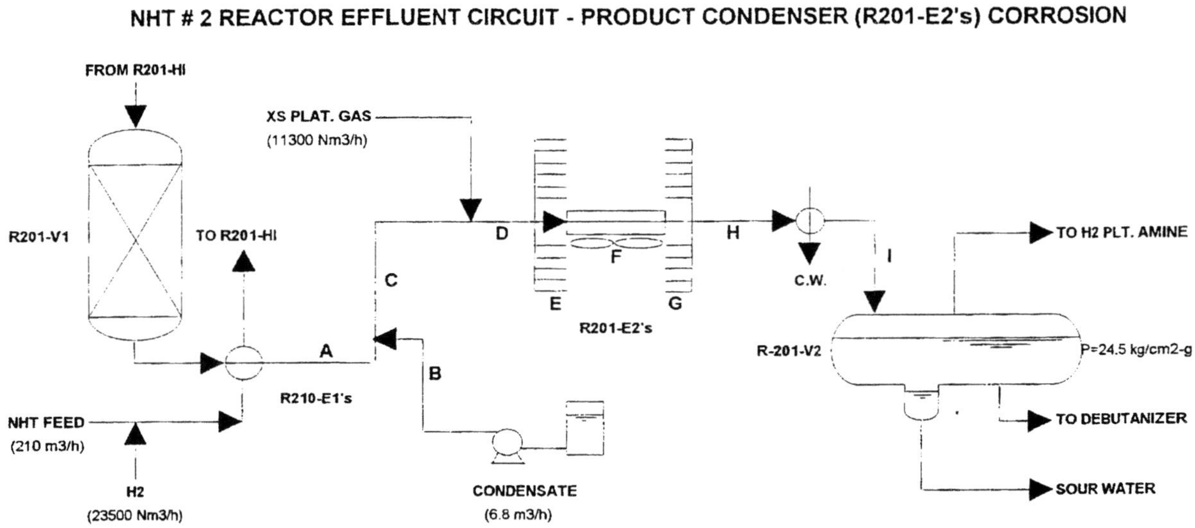

LOCATION	A	B	C	D	E	F	G	H	I
VELOCITY (m/s)	8.5	1.1	8.9	11.7	2.2	1.6 - 0.1	1.6	8.5	7.2
CORROSION (mpy)	N/A	N/A	30 - 60	30 - 80	30 - 100	15 - 25	N/A	N/A	N/A
TEMPERATURE (C)	186	45	144	138	138	138 - 88	88	88	46

Figure 5.8 - Process flow diagram

CASE HISTORY 5.8

Group/Type: ferrous, ASTM A106 Gr. B Sch. 80 carbon steel forging
Phenomenon: erosion-corrosion
Equipment: air-cooled condenser inlet nozzle
Environment: hydrotreater naptha
Conditions:

a) major species: naptha, hydrogen
b) minor species: NH_4HS, H_2S, water, Cl- up to 500 ppm
c) temperature: 290°F (138ºC)
d) pressure: 24.5 kg/cm^2
e) aeration: 1-2 ppm O_2 in wash water

Time to failure: 5 years

Illustration: process flow diagram
Comment: Fouled heat exchangers upstream of this air cooler led to higher than normal process temperatures. The unexpectedly high temperatures vaporized the wash water at higher than desirable rates. Consequently, the ammonium bisulfide concentration of the first fluids to condense was much higher than normal, leading to erosion-corrosion of the inlet nozzles at rates up to 136 mils per year (3.4 mm/y). The corrosion was accelerated by chlorides carried in from the hydrotreater catalyst and by an undesirably high oxygen content in the wash water.
Remedy: Increase wash water capacity to a minimum of 5% of total feed rate, clean the fouled upstream heat exhanger, and reduce wash water O_2 content to <20 ppb.

CASE HISTORY 5.09

Phenomenon: dew point erosion
Group/ Type: SS, 17-4 PH SS, condition H1150, HRC 32 (UNS S17400)
Equipment: gas compressor impellor
Environment: process gas stream
Conditions:
- (a) major species: ethyl ene, hydrogen, car bon monoxide and organic solvents
- (b) minor species: not given
- (c) temperature: 95°F (35°C)
- (d) pressure: not given
- (e) aeration: anaerobic

Time to failure: 3 months
Illustration: photomacrograph
Comment: A gas compressor operated at a speed of 23,000 rpm, moving 500 to 600 mcf/h of gas at dew point temperature. After three months in service, the 0.0625-in. nominal thickness of the impeller vanes had been reduced 50 to 85% along the outer edges. The vanes were still original thickness near the hub, where the gas velocity is lowest. The gas mixture by itself is non-corrosive. However, at its dew point, any gas can form liquid droplets, which, if impelled with sufficient velocity, will erode even the hardest materials.
Remedy: Raise the temperature of the inlet gas well above the dewpoint so droplets will not form.

CASE HISTORY 5.10

Phenomenon: fretting
Group/ Type: copper alloy/nickel-aluminum bronze, CA 955 (UNS C95500)
Equipment: intensifier head wave ring
Environment: atmospheric between steel heads
Conditions:
(a) major species: air
(b) minor species: water vapor
(c) temperature: ambient to -40°F (-40°C)
(d) pressure: 40,000 psi internal
(e) aeration: saturated
Time to failure: 2 months (2.6 million cycles)
Illustration: macro
Comment: This nickel-aluminum bronze wave ring sealed the head on a high pressure gas intensifier. After two months in continuous service, the wave ring began leaking. On examination, extensive fretting, spalling, and surface cracking was observed on the seating shoulders of the wave ring. Aluminum bronze is usually quite compatible with hardened steel for metal-to-metal pairs. The investigation revealed that the fundamental reason for the fretting was relative motion between the intensifer head and the cylinder body, resulting from different elastic responses to internal pressure.
Remedy: The head design was changed to reduce its stiffness and thus reduce the relative motion between head and cylinder.

CASE HISTORY 5.11

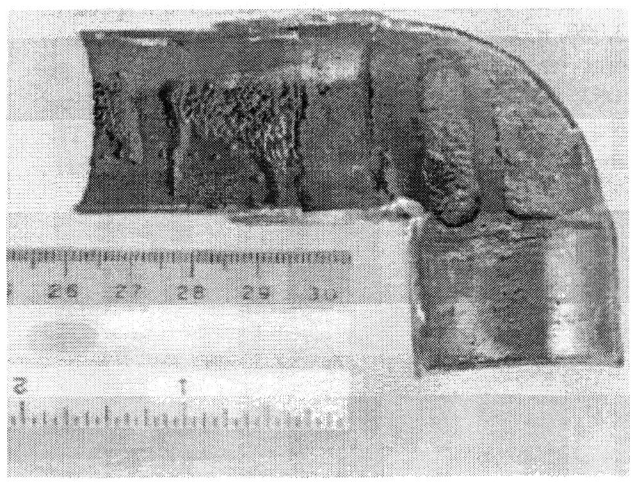

Phenomenon: erosion corrosion
Group/ Type: copper alloys, DHP copper (UNS C12200)
Equipment: commercial hot water piping
Environment: domestic hot water
Conditions:
(a) major species: Houston, TX, city water
(b) minor species: 200 ppm calcium carbonate alkalinity
(c) temperature: 145°F (63°C)
(d) pressure: 35 psi
(e) aeration: 3 to 5 ppm O_2
Time to failure: 2.5 years
Illustration: macro
Comment: Copper is widely used for domestic hot water piping since it is less prone to corrosion than bare or galvanized steel. However, it is more sensitive to velocity. This elbow is from the copper hot water system in a commercial hotel. An expansion of the system left the piping undersized for the flow rates. Flow in excess of 5ft/s was forced through these pipes. Resulting leaks showed the classic dimpled, rippled appearance of velocity-related corrosion.
Remedy: Reduce the velocity to below 3 ft/s in copper piping.

CASE HISTORY 5.12

Phenomenon: cavitation corrosion
Group/Type: nickel-based alloys, alloy 625 (UNS N06625)
Equipment: flow control valve
Environment: desalting waste water
Conditions:

a) major species: water, NaCl
b) minor species: H_2S, CO_2
c) temperature: 100 to 160°F (37.8 to 71°C)
d) pressure: 3,415 psia inlet, flow 800 US gal/min.
e) aeration: deaerated

Time to failure: 6 years
Illustration: macro
Comment: Despite the Stellite 6 (UNS R3006) hardfaced layer applied to the alloy 625 (UNS N06625) plug, the corroded zone had an appearance typical of cavitation corrosion. Hardness was low in the hardfaced layer, and x-ray fluorescence analysis confirmed that the hardfaced layer had suffered heavy weld dilution. The reduced hardness had reduced the cavitation resistance.
Remedy: Use a two-pass weld overlay procedure with Stellite 1 (R30001) or Stellite 12 (R30012) hardfacing alloy to minimize weld dilution and maximize hardness in the hardfaced layer.

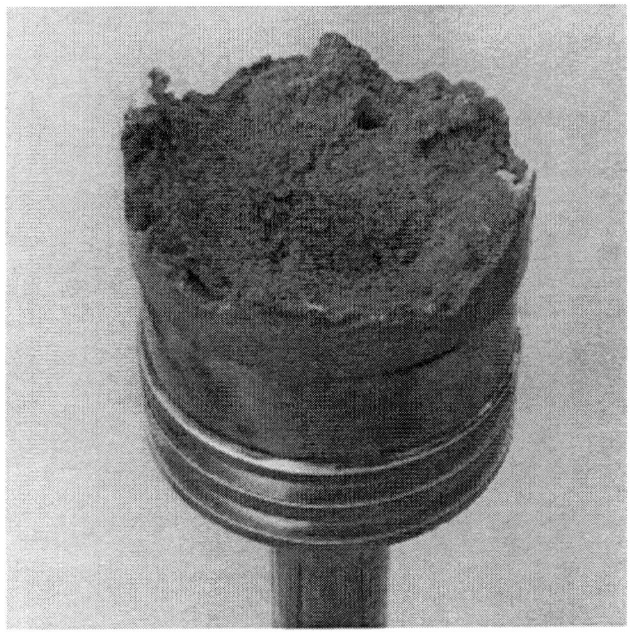

CHAPTER 6

Intergranular Corrosion

CASE HISTORY 6.1

Phenomenon: exfoliation corrosion
Group/ Type: light metals, aluminum, AA 7075-T6 (UNS A97075)
Equipment: aircraft landing gear brace
Environment: marine atmosphere
Conditions:
(a) major species: oxygen, nitrogen
(b) minor species: water vapor, chlorides
(c) temperature: ambient
(d) pressure: atmospheric
(e) aeration: saturated
Time to failure: not given
Illustration: macro, photomicrograph
Comments: AA 7075 in the -T6 condition is particularly susceptible to IGA in the marine atmospheres in which many aircraft must operate. When the part is machined such that transverse grain structure is exposed to the environment, yet the principal stresses are parallel to the longitudinal direction, exfoliation can occur. Exfoliation takes the form of leafy swelling caused by the accumulation of corrosion products along the elongated grain boundaries. On metallographic examination, the fundamentally intergranular nature of the attack can be seen.
Remedy: Overage to the -T73 condition if strength requirements permit. Anodize, paint, and seal fastener holes.

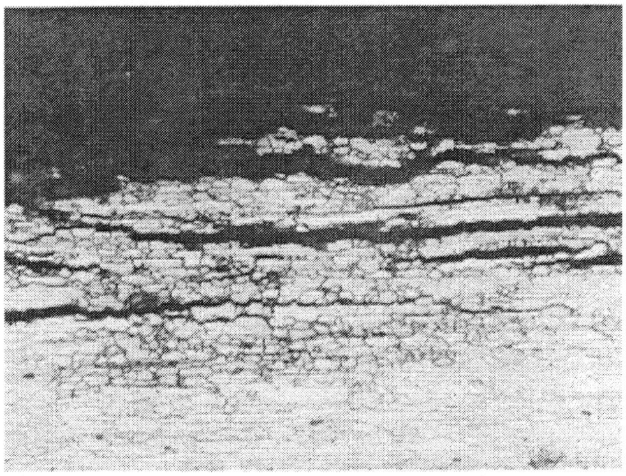

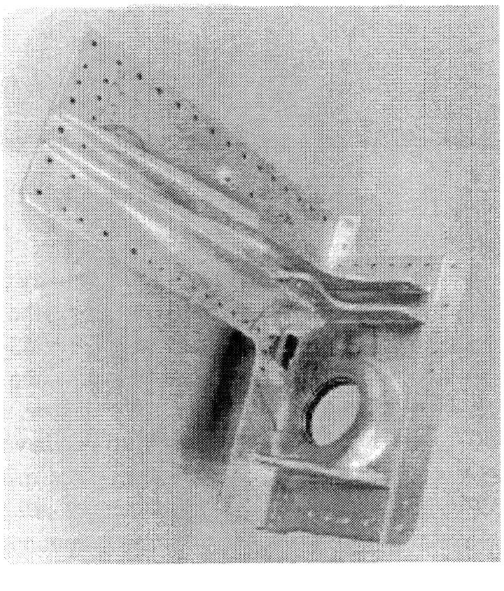

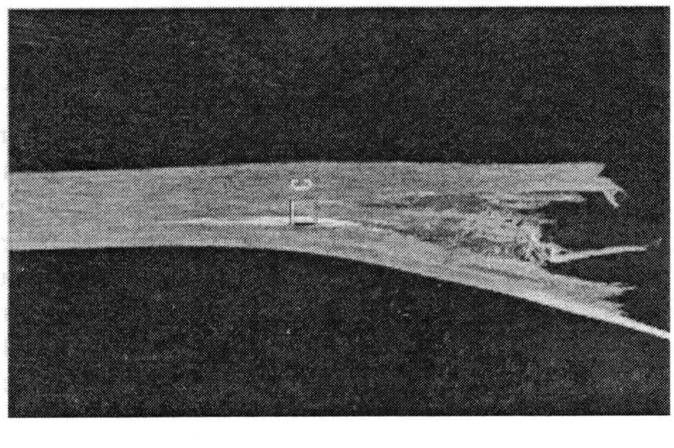

CASE HISTORY 6.2

Phenomenon: exfoliation
Group/Type: aluminum alloy, Alclad 7075-T6 (UNS A97075)
Equipment: aircraft wing skin
Environment: seacoast atmosphere, 4 km from the Arabian Gulf
Conditions:

a) major species: air
b) minor species: relative humidity 40 to 100%, airborne salt, blowing sand
c) temperature: Arabian Gulf ambient, 32 to 120ºF (0 to 49°C)
d) pressure: 1 atmosphere
e) aeration: 20%

Time to failure: 15 years
Illustration: macro, micro
Comment: This aircraft had been repainted one year before the damage was discovered, using a zinc chromate wash primer and a white polyurethane topcoat. Although it operated from an airstrip within sight of the Arabian Gulf, the aircraft was regularly washed with a slightly alkaline liquid soap to remove accumulated salt. One year after repainting, with 15 years total service on the aircraft, the Alclad layer was completely consumed on the underside of the wing skin on this cargo aircraft. Near the fastener holes, severe exfoliation corrosion had reduced the effective thickness of the wing skin to approximately 35% of original thickness. Examination indicated that the last paint job had been applied over a surface which had not been completely cleaned of old paint and corrosion products.
Remedy: Repair with -T6 material which has been given a retrogression and re-aging (RRA) heat treatment to eliminate exfoliation. Clean surfaces down to bare metal before repainting. Replace the alkaline soap with a neutral soap with 5% sodium chromate corrosion inhibitor additions.

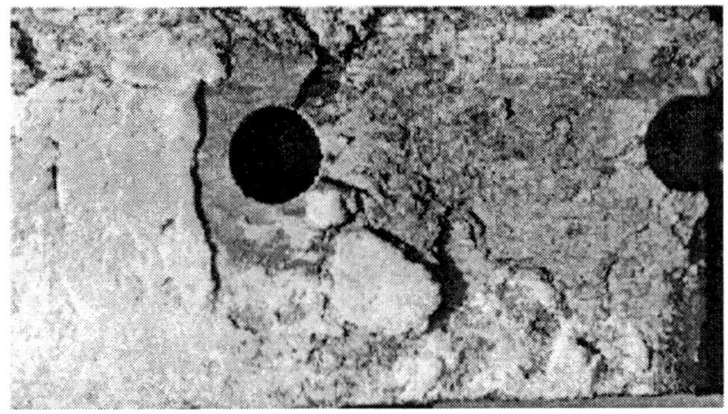

Figure 6.2(a) - Macrograph of damaged wing skin.

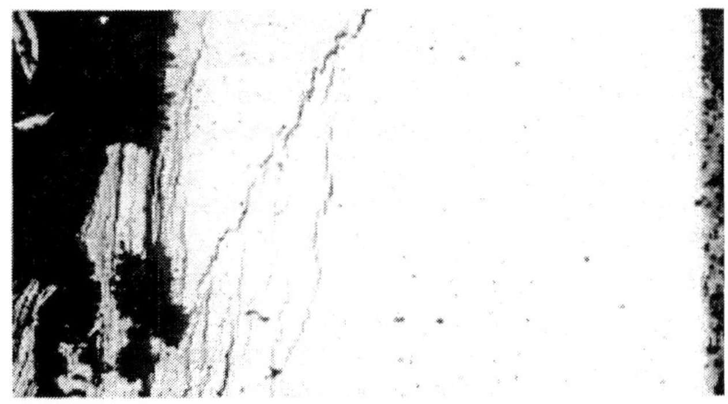

Figure 6.2(b) - Cross-section through damaged wing skin.

CASE HISTORY 6.3

Phenomenon: IGSCC
Group/ Type: light metals, aluminum alloys, AA 6063-T6 (UNS 96063)
Equipment: cooling tower fan blades
Environment: cooling tower mist
Conditions:

(a) major species: air
(b) minor species: water vapor
(c) temperature: ambient dew point
(d) pressure: atmospheric
(e) aeration: saturated

Time to failure: 7 years
Illustration: micro

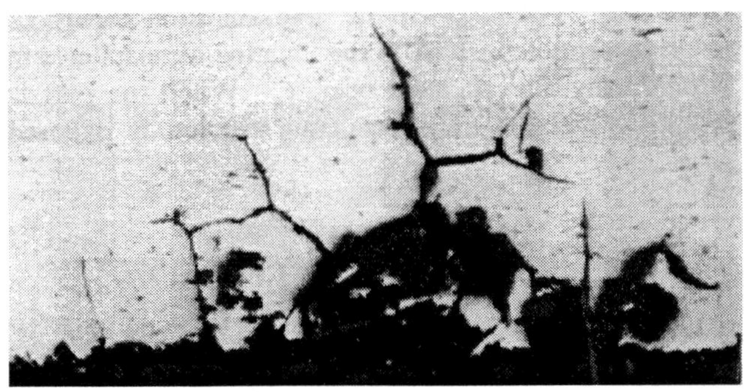

Figure 6.3 - Micro

Comments: The 6000 series alloys are normally considered immune to the SCC and IGA that plague the higher strength aluminum alloys. In this case, AA 6063-T6 alloy extrusions were welded into an assembly, then stress-relieved. Inspected after 7seven years in service, numerous short, shallow cracks were observed just beyond the end of the welded external stiffener. The path was clearly intergranular. The effect of the stress-relief treatment on possible sensitization was not determined.
Remedy: Replace with fiberglass fan blades.

CASE HISTORY 6.4

Phenomenon: IGA
Group/ Type: light metals, aluminum alloy, AA 6061-T6 (UNS A96061)
Equipment: rocket-fuel tank bellows
Environment: pickling and cleaning solution
Conditions:
(a) major species: water, chromic acid
(b) minor species: chlorides
(c) temperature: ambient
(d) pressure: 1 atm
(e) aeration: saturated
Time to failure: 1 week
Illustrations: SEM and optical micrographs
Comment: This rocket engine fuel tank bellows leaked on pressure test. On examination, through-thickness leak paths caused by intergranular corrosion were found initiating at the bottom of each bellows convolution. Since the bellows had not yet seen service, the cause of corrosion was believed to be residue of a chromic acid cleaning solution sitting in the bellows. EDX analysis revealed a high surface chloride ion concentration so bath contamination with choride-bearing water was suspected.
Remedy: Make up pickling solutions with distilled water and thoroughly dry the bellows after pickling.

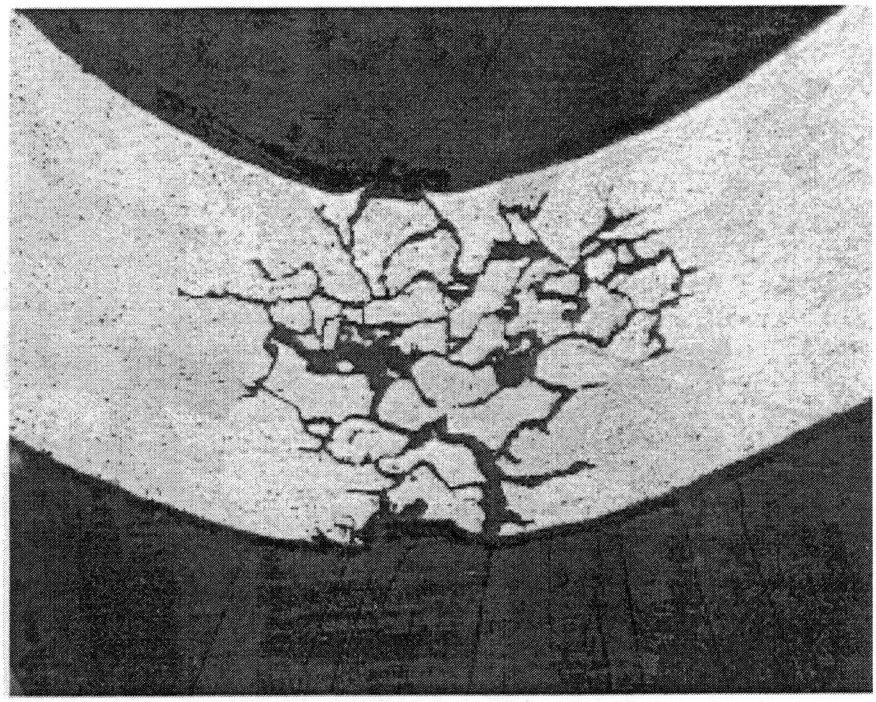

Figure 6.4(a) - Micro of cross-section, 50X.

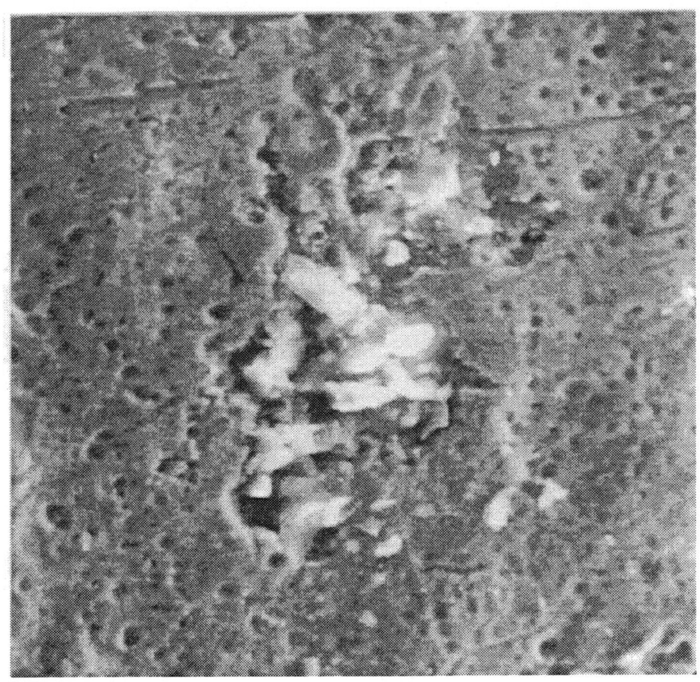

Figure 6.4(b) - SEM micro, 300X.

CASE HISTORY 6.5

Phenomenon: IGA (followed by corrosion fatigue)
Group/ Type: ferrous, AISI 403 SS (UNS S40300)
Equipment: air compressor
Environment: plant air
Conditions:

(a) major species: air
(b) minor species: water vapor, ammonium sulfate, ammonium phosphate, nitric acid fumes
(c) temperature: ambient (inlet)
(d) pressure: atmospheric (inlet)
(e) aeration: 20%

Time to failure: not given
Illustration: macro, SEM fractograph
Comments: Compressor blades were coated with an aluminum-rich, high-temperature coating for protection against corrosive fumes in the plant atmosphere. However, after the coating was consumed, pitting and IGA began. On the tension sides of the blades, fatigue cracks initiated at these intergranular pits. Eventual blade failure with extensive compressor damage resulted.
Remedy: Inspect and re-coat blades yearly.
Reference: K.F. Krysiak, *Materials Performance*, 22, 11, (1983): pp. 23-28.

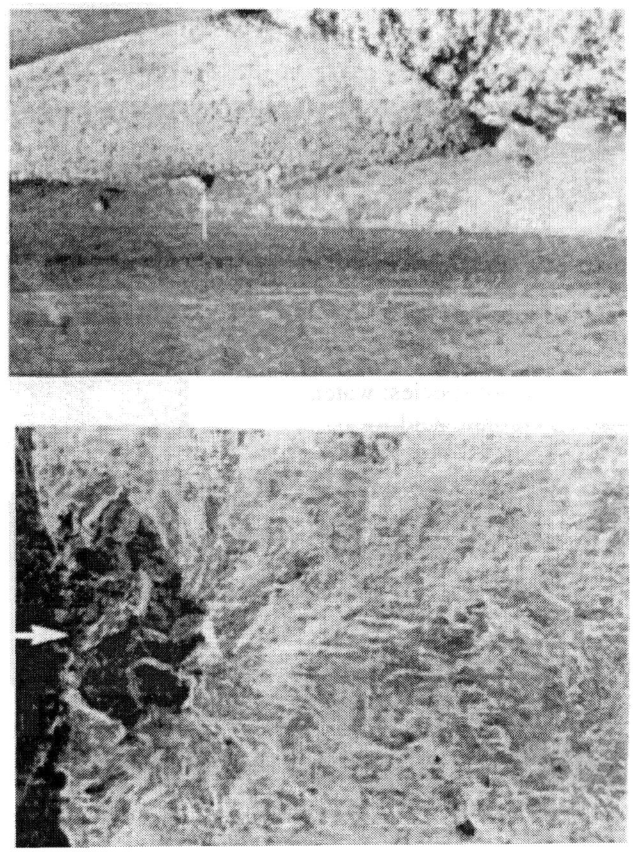

Figure 6.5 (a) and (b)

CASE HISTORY 6.6

Phenomenon: intergranular attack
Group/Type: ferrous, AISI type 304 SS (UNS S30400)
Equipment: cavitation control device on a firewater dump valve
Environment: marine atmosphere, Arabian Gulf seacoast
Conditions:

a) major species: air
b) minor species: water vapor, salt spray
c) temperature: Arabian Gulf ambient (50-120ºF)
d) pressure: normally atmospheric
e) aeration: 20%

Time to failure: unknown
Illustration: photograph, photomicrograph
Comment: This cavitation control device consisted of a number of type 304 stainless steel tubes brazed together on the atmospheric side of a pressure relief valve. Normal environment is seacoast air, with occasional exposure to seawater when the relief valve opened. After failure, metallographic examination showed that the heat of brazing had sensitized the tubing, leaving it vulnerable to intergranular attack.
Remedy: Replace with AISI type 316L stainless steel, which will not sensitize during the brazing process.

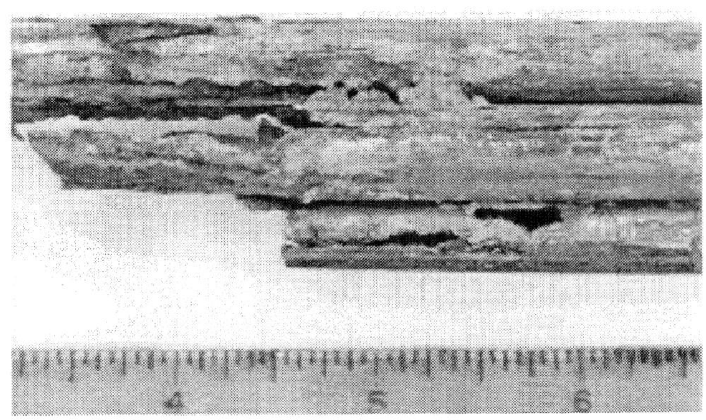

Figure 6.6(a) - Photograph

CASE HISTORY 6.7

Phenomenon: intergranular corrosion
Group/ Type: martensitic SS
Equipment: surgical scissors
Environment: body fluids, distilled water, steam
Conditions:
(a) major species: distilled water, body fluids
(b) minor species: NaCl, trace elements
(c) temperature: ambient to 97°F (36.1°C), 250°F (121°C) during sterilization
(d) pressure: 1 atm
(e) aeration: saturated

Time to failure: 4 months of dry storage, 6-12 operations
Illustration: macro
Comment: Even low corrosion rates cannot be tolerated in surgical instruments since the sterility of instruments showing corrosion products always will be suspect. These surgical scissors were exposed alternately to body fluids and steam sterilization during half a dozen surgical operations. The scissors had a wear-resistant tungsten carbide blade brazed onto the hardened martensitic SS body. The brazing operation was apparently conducted at temperatures in the chromium carbide precipitation range; massive carbide particles are visible in the microstructure near the braze joint. A thin border of preferential corrosion developed at the chromium-depleted zone near the braze joint.

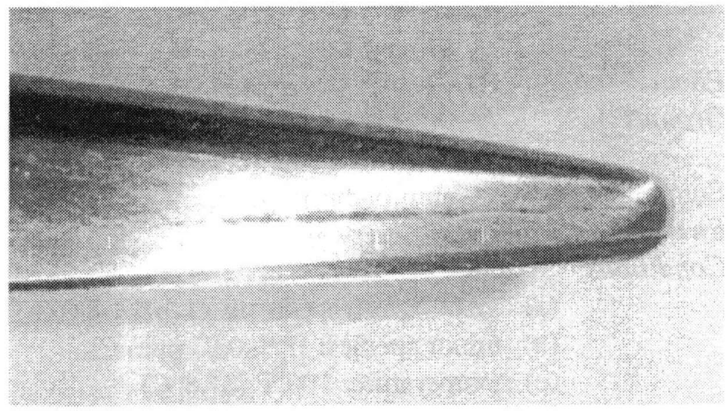

Proposed remedy: Select a solder or braze alloy or joining method that can be applied at temperatures below the chromium carbide precipitation range.

CASE HISTORY 6.8

Phenomenon: end grain attack
Group/Type: ferrous, AISI 420 stainless steel (UNS S42000)
Equipment: pump impeller lock nut
Environment: wash water
Conditions:
a) major species: water
b) minor species: 1,273 mg/L Na+, 3,080 mg/L Cl-
c) temperature: Arabian Gulf ambient
d) pressure: 300 psig
e) aeration: 60 ppb dissolved O_2

Time to failure: 3 years
Illustration: photograph
Comment: These nuts were machined from bar stock and used to lock CA6NM pump impellers in place on the pump shaft. After three years in service the freely exposed outer surface of the AISI 420 nuts showed no attack. However, the faces, where an end grain microstructure was exposed to a crevice, showed heavy pits up to ¼-inch deep following sulfide stringers in the microstructure.
Remedy: Replace with CA6NM nuts
Contributor: S.B. Jones, Saudi Aramco, Dhahran, Saudi Arabia.

Figure 6.6(b) - Photomicrograph

CASE HISTORY 6.9

Phenomenon: IGA causing corrosion fatigue
Group/Type: ACI CF-8M SS (UNS J92900) [cast AISI 316 SS (UNS S31600)]
Equipment: vacuum pump impeller
Environment: sour refinery gas
Conditions:
(a) major species: SO_2, H_2O (pH 1.8)
(b) minor species: H_2S, 0.03 ppm Cl-
(c) temperature: 100°F (37.8°C)
(d) pressure: vacuum
(e) aeration: anaerobic

Time to failure: 2.5 days at 1,750 rpm
Illustration: macro, micro
Comment: Carbon from organic binders in the mold contaminated the surface of this cast SS impeller, leading to IGA. Fatigue cracks began at the IGA, producing rapid catastrophic failure.
Remedy: Specify silica flour coating on the inside of the mold for SS castings. Test for solution-anneal after shakeout using ASTM A262 method A.

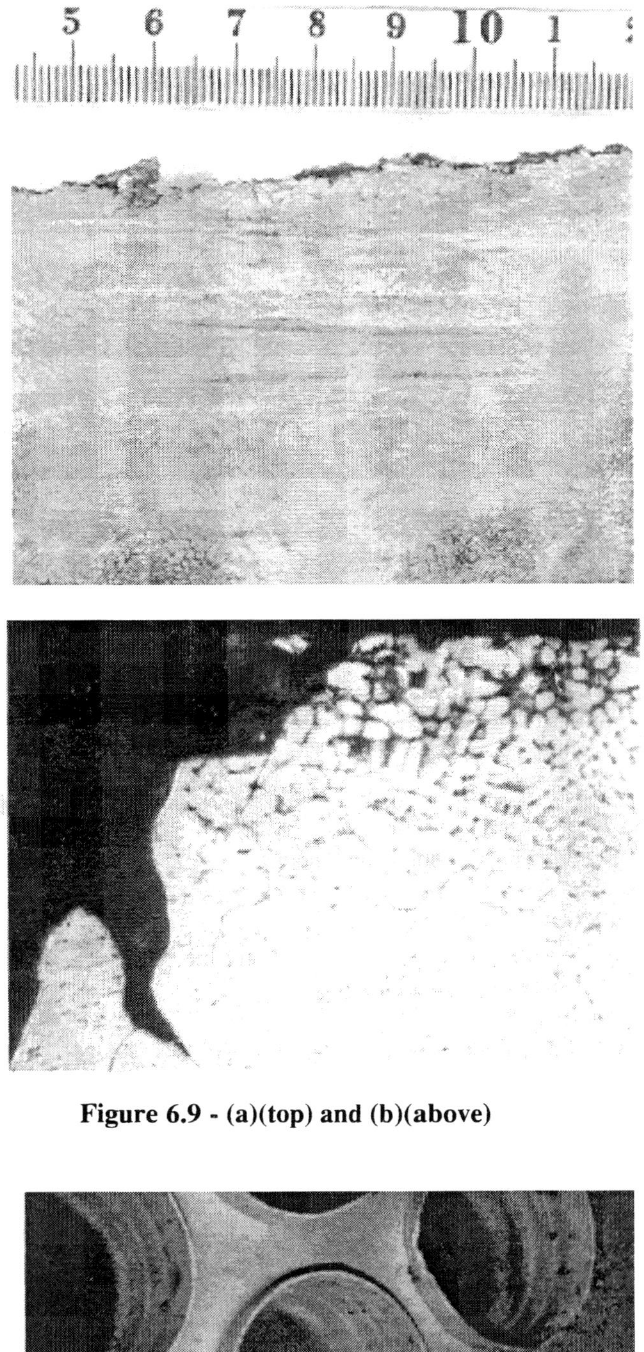

Figure 6.9 - (a)(top) and (b)(above)

CASE HISTORY 6.10

Phenomenon: IGA
Group/ Type: SS, ASTM A268 grade XM-27 (UNS S44625)(26-1 low-interstitial ferritic)
Equipment: heat exchanger tubes
Environment: caustic
Conditions:
(a) major species: 42% NaOH, 10% Na_2SO_4, 7% NaCl in H_2O
(b) minor species: 0.2% Na_2CO_3, 0.1% $NaClO_3$
(c) temperature: 325 to 340°F (162.7 to 171°C)
(d) pressure: 50 psig
(e) aeration: not given

Time to failure: 5 years
Illustration: macro, micro (pages 102 and 103)
Comment: The 26Cr-1Mo SS tubes in this heat exchanger lost thickness from IGA at the rate of ~ 16 mils/y (0.4 mm/y) until the tube walls were too thin to resist the pressure on the shell (steam) side. Investigation revealed that the tubes were from a lot without columbium, whose carbon and nitrogen levels exceeded specifications. As a result both the seam welds and the base metal were sensitive to IGA after quench annealing.
Remedy: Use tubing with columbium additions to ASTM requirements (Nb 7.7(%C)+0.034). Test incoming tubing lots with ASTM A262, practice C, the 65% nitric acid corrosion test).

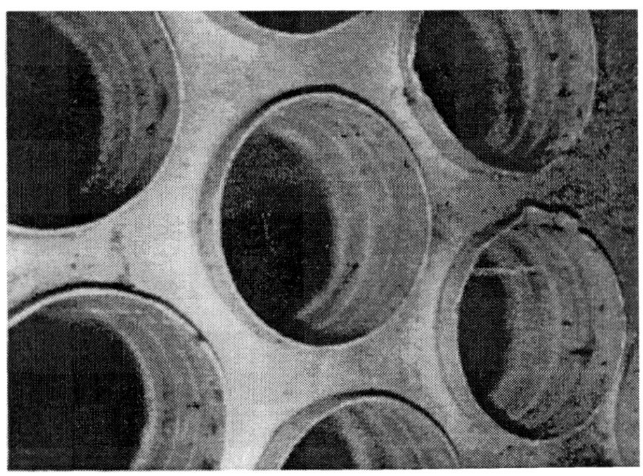

Figure 6.10(a) - Macro

CASE HISTORY 6.11

Phenomenon: intergranular attack
Group/Type: copper-based alloys, aluminum brass UNS C68700
Equipment: seawater evaporator tubes
Environment: seawater, steam condensate, sulfuric acid
Conditions:

a) major species: water, 3.5% NaCl, followed by H_2SO_4
b) minor species: Mg, Ca, Br ions in Red Sea seawater
c) temperature: 107ºC
d) pressure: 0.35 kg/cm2
e) aeration: saturated

Time to failure: six months total
Illustration: micro

Comments: This device made freshwater by using 5 psig steam to evaporate seawater. Sulfuric acid was added to the seawater to reduce scale deposited inside the tubes. After six months service, a hydrotest revealed that 297 tubes out of a total of 4,400 were leaking. Metallographic examination revealed extensive intergranular attack and SCC.

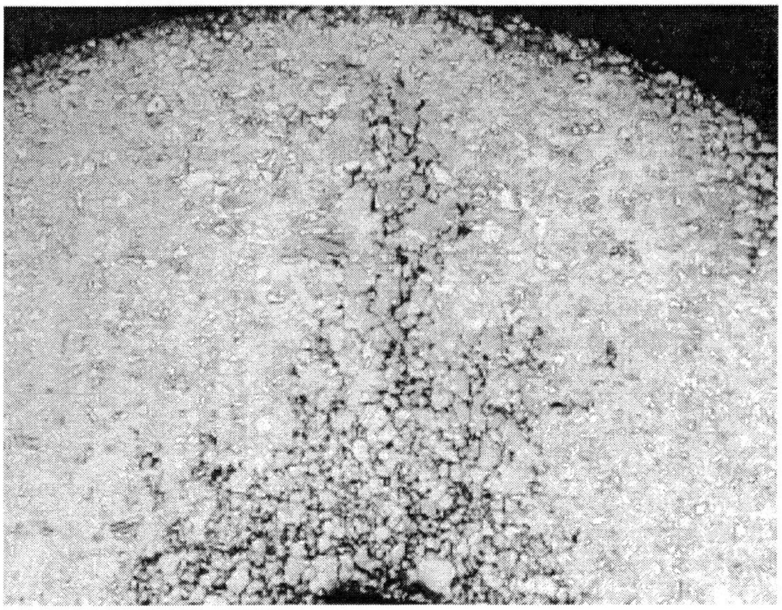

Figure 6.11 - Photomicrograph of cross section, showing IGA of Al brass tube

Remedy: Switch to an ambient-temperature reverse-osmosis process for desalinating the seawater.

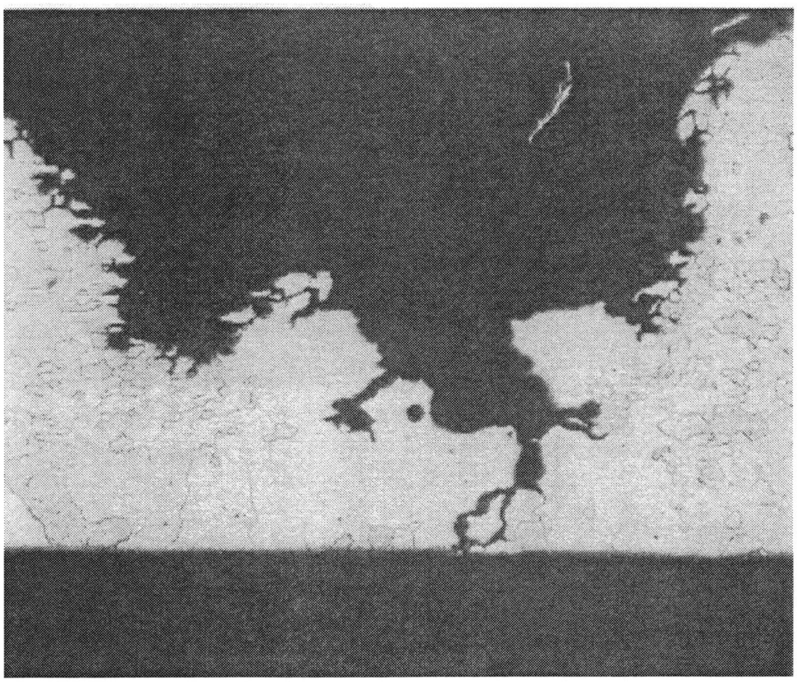

Figure 6.10(b) - Micro

CASE HISTORY 6.12

Phenomenon: intergranular attack
Group/Type: nickel-based alloys, Waspaloy (UNS N07001)
Equipment: fasteners on a gas turbine
Environment: sulfur-bearing combustion gases
Conditions:

a) major species: carbon dioxide, carbon monoxide
b) minor species: 0.72% H_2S
c) temperature: 1,112ºF (600°C) in combustion section
d) pressure: 1 atmosphere
e) aeration: none

Time to failure: 9,185 hours
Illustration: fracture surface macro, SEM fractograph, micro
Comment: Fasteners attaching the inter turbine duct to the first stage power turbine stator casing were intended for exposure only to dry air at an estimated 400ºF (204°C). However, in this installation, leakage of hot combustion gases out of the duct resulted in fracture of 35 out of 90 fasteners. A heavy black sulfide surface scale formed on the fasteners; intergranular fractures resulted, with elemental sulfur deposited in the cracks (micro).
Remedy: Eliminate leakage of the combustion gases from the duct onto the fasteners.

Figure 6.12(a) - Fracture surface

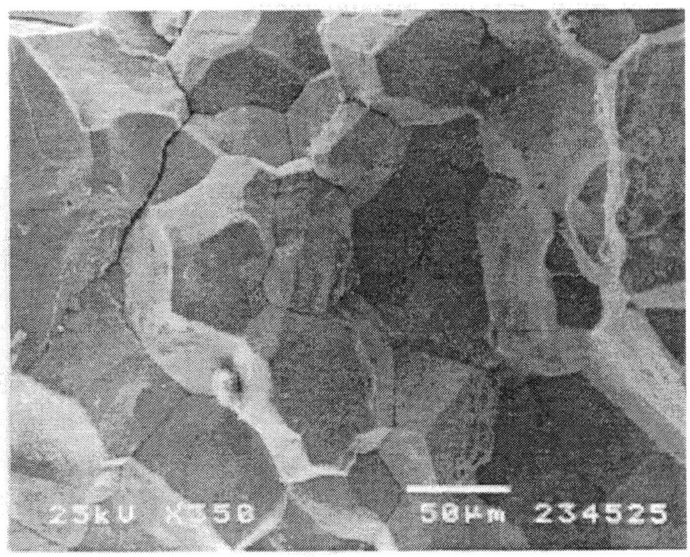

Figure 6.12(b) - SEM fractograph

Figure 6.12(c) - Photomicrograph

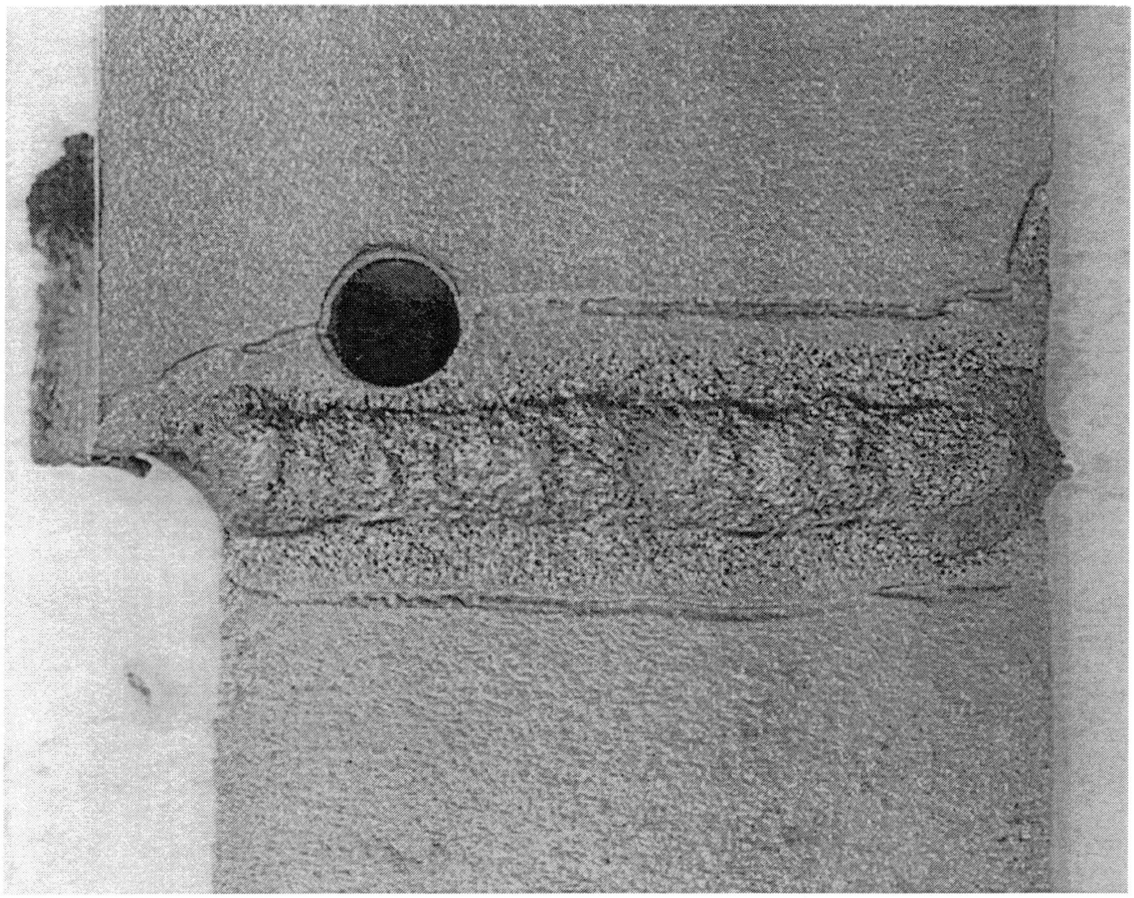

CASE HISTORY 6.13

Phenomenon: IGA in welds and HAZ
Group/ Type: zirconium alloy, type 702 (UNS R60702)
Equipment: distillation tray stiffener
Environment: sulfuric acid
Conditions:
(a) major species: 56% H_2SO_4
(b) minor species: 393 ppm Fe, 3 ppm Cu, 510 ppm Cl-
(c) temperature: 248°F (120°C)
(d) pressure: vacuum
(e) aeration: none

Time to failure: 10 years
Illustration: macro
Comment: Both the distillation tray and its stiffener exhibited preferential attack in arc welds and HAZs. The welds corroded along alpha-prime needles. The HAZs showed IGA. The preferential attack is attributed to grain boundary precipitates formed during the relatively high heat input of arc welding. Nearby spot welds (with their lower heat input) were not attacked.
Remedy: Heat treat at 1,350°F (732°C) in air to redissolve the grain boundary precipitates and repair the oxide film.

CHAPTER 7

Dealloying

CASE HISTORY 7.1

Phenomenon: dealloying
Group/Type: nickel-bearing cast iron, Type 1B Ni-resist
Equipment: pump impeller
Environment: underground aquifer water
Conditions:

a) major species: water
b) minor species: 18,000 ppm Cl- ion, Na, Mg, sulfates
c) temperature: ambient
d) pressure: 300 psig
e) aeration: once per week

Time to failure: 10 years
Illustration: photomicrograph
Comment: Nickel-bearing cast iron once was widely specified for pump impellers because of its good corrosion resistance in freshwaters. However, in high-chloride underground aquifer water, it proved unsatisfactory. In this case, accelerated attack developed along the graphite particles (see micro), leaving a weak, porous surface layer. Strength loss eventually led to fracture of the impeller by corrosion fatigue.
Remedy: Replace with CF3M stainless steel impellers.

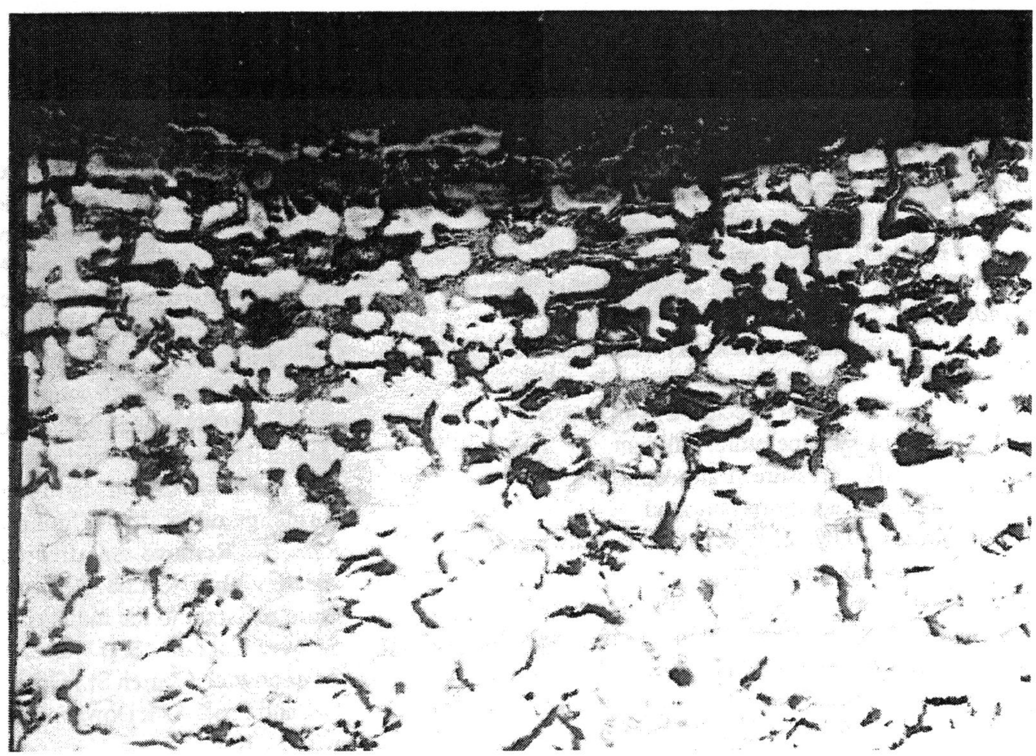

Figure 7.1 - Photomicrograph

CASE HISTORY 7.2

Phenomenon: dealloying
Group/Type: ferrous, gray cast iron
Equipment: air handling drainage piping
Environment: water condensed from air conditioners, miscellaneous drainage
Conditions:
a) major species: water, pH ~ 6
b) minor species: carbonates, traces of Ag_2S
c) temperature: room temperature
d) pressure: gravity
e) aeration: saturated

Time to failure: 10 years
Illustration: photograph, photomicrograph
Comment: This gray cast iron drainage pipe carried water condensed from air conditioners cooling a supercomputer as well as miscellaneous drainage, including that from a photo lab. After 10 years service the pipe developed seeping leaks and sizeable tubercles on the outside surface (see photo). Metallography indicated that the leaks occurred in areas of severe graphitization (photomicrograph), where the ferrite matrix had been completely consumed, leaving only graphite particles behind.
Remedy: Replace in kind, and reroute the drain from the photo lab to sewer in a separate PVC piping system.
Contributor: S.B. Jones, Saudi Aramco, Dhahran, Saudi Arabia

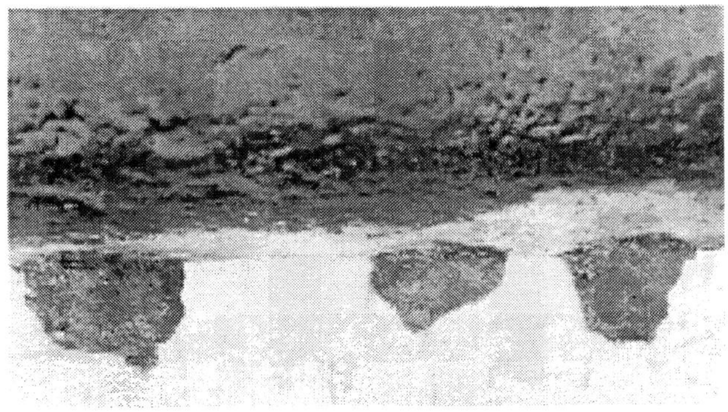

Figure 7.2(a) - Photo of tubercles on outside of leaking drainage pipe.

Figure 7.2(b) - Photomicrograph of section through a leak.

CASE HISTORY 7.3

Phenomenon: dealloying
Group/Type: copper alloy (UNS S28000)
Equipment: anti-foulant sheathing on wooden-hulled sailing vessel
Environments: 1) seawater, 2) industrial atmosphere
Conditions:
a) major species: 1) water with 3½ % NaCl, then 2) water vapor, air
b) minor species: 2) sulfates, SO_2
c) temperature: ambient
d) pressure: 1 atmosphere
e) aeration: saturated

Time to failure: 93 years total (80 years seawater immersion, 13 years industrial atm.)
Illustration: none
Comment: UNS C28000 sheathing effectively stopped marine biofouling on the wooden hull of the sailing vessel *Cutty Sark* during her 80 year career as a tea clipper, general cargo and sail training ship. However, once the ship was dry-docked as a museum exhibit, the sheathing began to bulge and split, sometimes crumbling to powder. Examination indicated heavy dezincification from the inner sheet surfaces outward. As the elm hull planking of the vessel dried and shrank, the inner surfaces of the copper-alloy sheets were exposed to the Thames River docklands atmosphere. Heavy concentrations of SO_2 in the air combined with sulfates leached from the wood to produce acidic conditions under the Muntz metal sheets. Dezincification followed, and the porous copper residue was too weak to withstand the mechanical stress due to shrinkage of the drying hull planks.
Remedy: Remove the affected Muntz metal sheets, resheath with UNS C68700, a copper-zinc-aluminum alloy more resistant to the industrial atmosphere.
Source: Captain S.T. Waite, M.N.I., master, *Cutty Sark*, 2 Greenwich Church St., Greenwich, London SE10 9BG, UK and Prof. A.G. Dowson, Royal Naval College, Greenwich, UK.

CASE HISTORY 7.4

Phenomenon: dealloying
Group/ Type: copper-nickel alloy 70-30 (UNS C71500)
Equipment: structural support
Environment: phosphoric acid
Conditions:
(a) major species: 55% phosphoric acid
(b) minor species: not specified
(c) temperature: 482°F (250°C)
(d) pressure: not given
(e) aeration: anaerobic
Time to failure: 57 days
Illustration: micro
Comment: After 57 days in service, a reddish, porous surface layer was observed on 70Cu30Ni welds in a phosphoric acid column. Investigation (Figure) showed denickelification to a maximum depth of 0.004 in., indicating a dealloying rate of 0.038 ipy in this service.
Remedy: Use phosphor bronze weld rod as replacement.

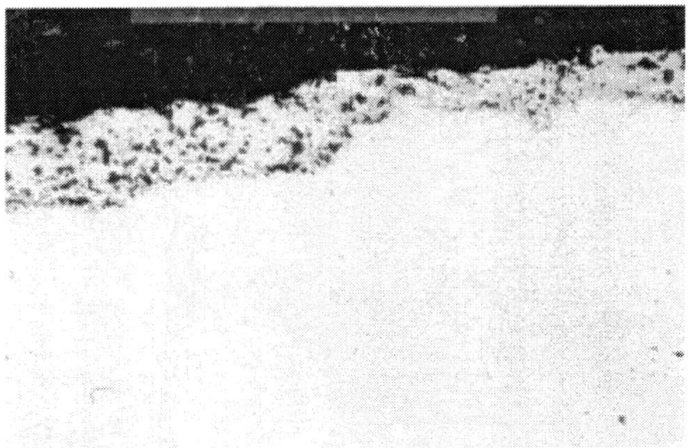

Figure 7.4 - Micro

CASE HISTORY 7.5

Phenomenon: selective leaching
Group/Type: nickel-based alloy, Cr-Fe-B-Si-Ni hardfacing
Equipment: pump shaft sleeve
Environment: Arabian Gulf seawater
Conditions:
a) major species: water, 5% NaCl
b) minor species: sulfates, carbonates, silica sand
c) temperature: ambient (55-100ºF)
d) pressure: 1 atm.
e) aeration: saturated
Time to failure: 9 months
Illustration: photomicrograph
Comment: A 316 stainless steel pump shaft sleeve had a flame-sprayed hardfacing with 15% Cr, 4% Fe, 76% Ni, balance boron and silicon. Seawater penetrated the voids in the flame-sprayed coating, setting up occluded cells which selectively leached out the chromium, iron, boron and silicon, leaving a relatively weak spongy mass of nickel which was prone to mechanical spalling and cracking.
Remedy: Switch to a seawater-resistant, higher-chromium, cobalt-based hardfacing applied by weld overlay over a UNS 20910 shaft sleeve.

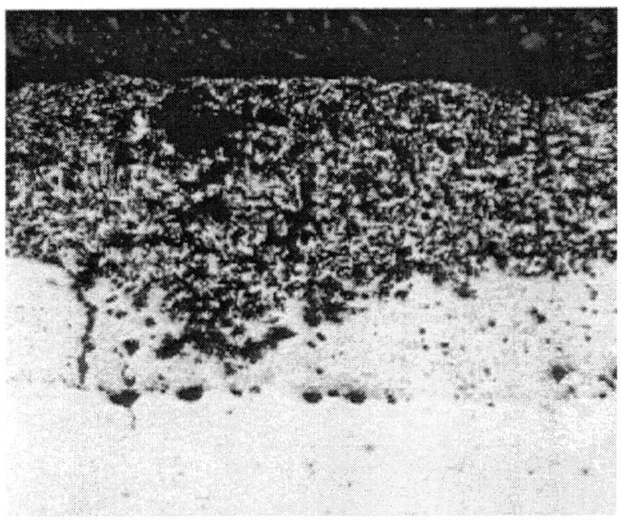

Figure 7.5 - Photomicrograph

CHAPTER 8

High Temperature Corrosion Performance

CASE HISTORY 8.1

Phenomenon: stress-rupture
Group/ Type: carbon-0.25% molybdenum steel
Equipment: heat exchanger tubing
Environment: tube side-hot hydrocarbon gas; shell side-steam
Conditions:
(a) major species: steam
(b) minor species:
(c) temperature: tube side 1,382°F (750°C); shell side 608°F (320°C)
(d) pressure: 1,700 psi (shell side)
(e) aeration: none
Time to failure: less than 48 hours (total service life seven years)
Illustration: micro
Comment: This heat exchanger cooled 1,382°F (750°C) hydrocarbon gas with 1,700 psi steam on the shell side. Power outages interrupted steam flow several times over a weekend. When steady-state operation was resumed, 7 out of the 10 innermost tubes had pulled loose from their welds in the tubesheet and were leaking. Metallography showed the decarburization and intergranular crack pattern typical of stress-rupture. It was concluded that the interruptions in steam flow had allowed the innermost tubes to become too hot, exceeding their stress-rupture temperature, despite the relatively short exposure time.
Remedy: Couple hot gas and steam flow so that gas is cut off during coolant outages.

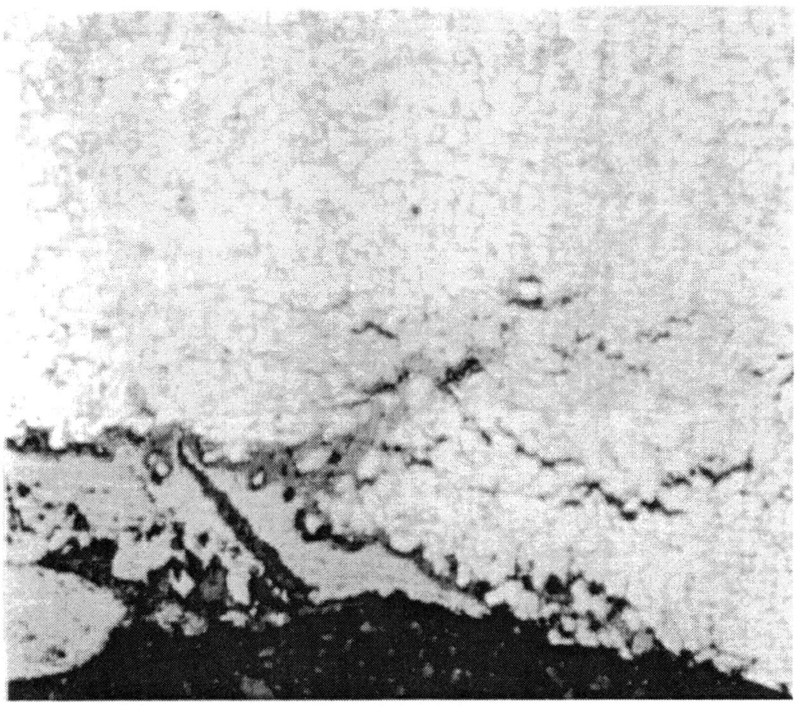

Figure 8.1 - Micro

Figure 8.2 - Catalyst support grating

CASE HISTORY 8.2

Phenomenon: sulfidation
Group/ Type: ferrous, carbon steel
Equipment: sulfur converter's catalyst support grating
Environment: sulfur converter
Conditions:

a) major species: 52% CO_2, 28.5% N_2
b) minor species: 6.5% H_2S, 3.5% SO_2, 9.3% H_2O
c) temperature: 700ºF (371°C) normal operating, 1,400ºF (760°C) for 1 h twice a year during regenerations
d) pressure: 15 psig
e) aeration: none

Time to failure: 33 months
Illustration: photograph
Comment: Catalyst regeneration temperatures in this unit were raised from 1,000ºF to 1,400ºF (538°C to 760°C) to improve efficiency. After less than three years, the grating was heavily scaled, buckled, and its thickness was reduced from 0.153 in. to 0.077 in., a corrosion rate of 0.027 in./y due to sulfidation.
Remedy: Replace with AISI 321 SS (UNS S32100).

CASE HISTORY 8.3

Phenomenon: high-temperature IGA
Group/ Type: ferrous, mild steel
Equipment: copper smelter
Environment: molten copper matte
Conditions:

(a) major species: copper, copper sulfide, iron sulfide, silica flux
(b) minor species: iron silicates, 5% sulfur dioxide, oxygen
(c) temperature: 2,192°F (1,200°C)
(d) pressure: atmospheric
(e) aeration: pressure blowing with oxygen-enriched air

Time to failure: less than 72 h
Illustration: photomicrograph
Comments: No commercially available metal alloy will survive the copper conversion environment, with its molten copper and sulfides, without rapid attack. This mild steel tuyere suffered deep intergranular penetration of the grain boundaries, with steel grains dropping out of the matrix and dissolving into the molten copper, within 72 h of installation.
Remedy: Use a combination of ceramics and metals to combat dissolution by the molten copper and thermal shock.

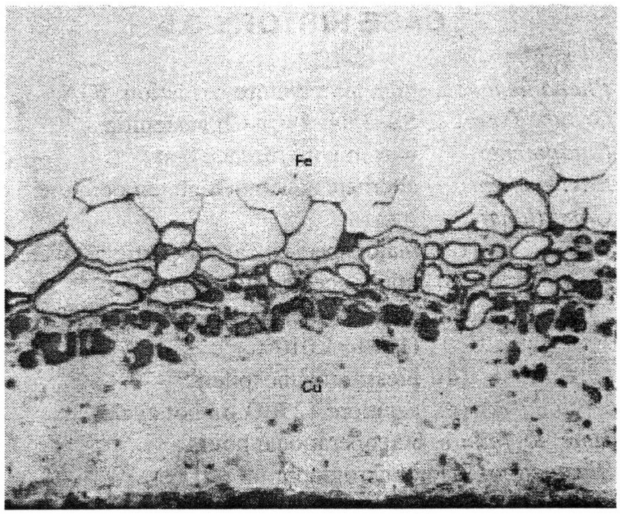

Reference: S.T. Pohlman, R.V. Olson, "Corrosion and Materials Problems in the Copper Production Industry," paper no. 229, CORROSION/84 (Houston, TX: NACE, 1984)

CASE HISTORY 8.4

Phenomenon: stress-rupture
Group/ Type: AISI 304 SS (UNS S30400)
Equipment: Economizer support bolts
Environment: flue gas
Conditions:

(a) major species: CO, CO_2, N_2
(b) minor species:
(c) temperature: 1,256°F (680°C)
(d) pressure: not recorded
(e) aeration: none

Time to failure: 2 months
Illustration: macro, micro
Comments: Illustrations show the rough, grainy, surprisingly brittle appearance of the fracture surface; little gross plastic deformation was observed. The photomicrograph shows evidence of internal fissures beginning at triple points (grain boundary intersections). The triple point fissuring is characteristic of high temperature stress-rupture.
Remedy: Replace with nickel-based alloy bolts such as alloy C-276 for better high temperature strength.

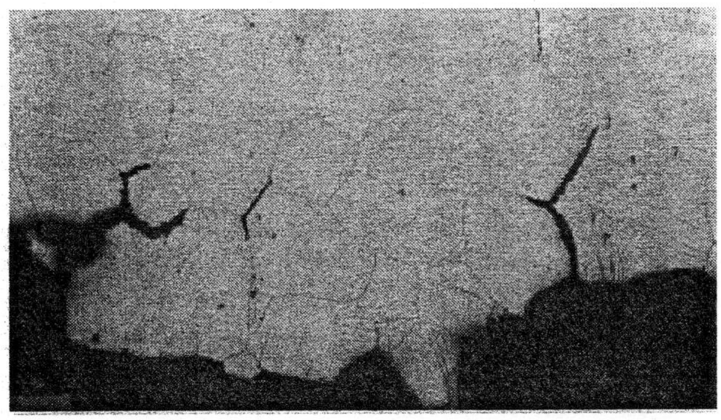

Figure 8.4(b) - (above) 150X

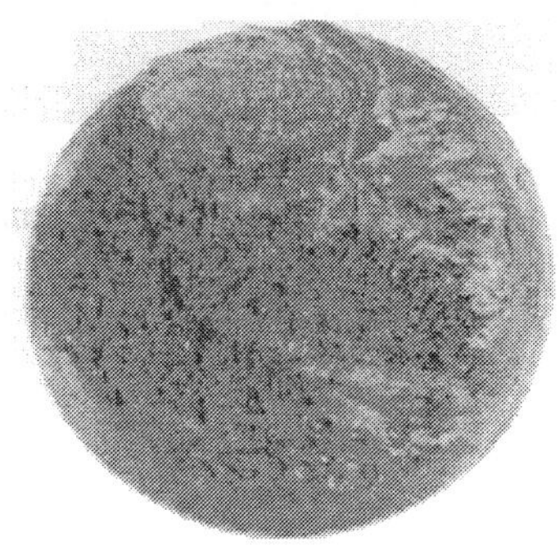

Figure 8.4(a) - (right) 3X

CASE HISTORY 8.5

Phenomenon: high temperature oxidation, IGA
Group/ Type: SS, 35Ni-19Cr-Cb austenitic
Equipment: woven wire furnace belt
Environment: alternate aqueous/high temperature
Conditions:
(a) major species: carbon particles, water
(b) minor species: 1,000 ppm Cl-
(c) temperature: cyclic 300 to 1,850°F (149 to 1,010°C)
(d) pressure: atmospheric
(e) aeration: 12% O_2 in hot cycle
Time to failure: 800 operational hours
Illustration: photomicrograph
Comments: This furnace regenerated wet, activated carbon particles by burning organic matter out of their pores. Wet carbon was fed onto an endless woven-wire belt that passed through a drying cycle at ~ 300°F (149°C), followed by a burnout cycle at 1,850°F (1,010°C). Twelve percent O_2 was added to the furnace atmosphere to aid combustion. The wire sensitized during the high temperature cycle. Aqueous IGA begins when the wet sludge first drops on the belt. The exposed grain boundaries then suffered accelerated oxidation during subsequent high temperature cycles.
Remedy: Switch to Al-containing low-carbon alloys.

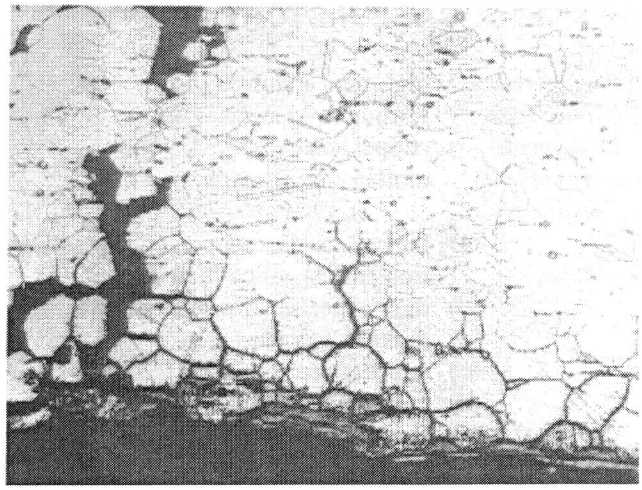

Figure 8.5 - Photomicrograph

CASE HISTORY 8.6

Phenomenon: hot salt corrosion
Group/ Type: SS, 35Ni-19Cr-Cb austenitic
Equipment: woven-wire furnace belt
Environment: waste incinerator
Conditions:
(a) major species: municipal sludge, water
(b) minor species: Na, P, S, O_2, Cl-
(c) temperature: cyclic, 300 to 1,700°F (149 to 927°C)
(d) pressure: atmospheric
(e) aeration: not given
Time to failure: 800 hours
Illustration: photomicrograph
Comments: The woven wire belts in this municipal sludge incinerator showed scattered areas of local attack. Since sodium, potassium, sulfur and oxygen are present in the sludge, the presence of Na_2SO_4 or K_2SO_4 salts was postulated. When molten, these salts flux the protective oxide off high temperature alloys, allowing rapid attack of the base metal. The presence of chloride will accelerate this attack.
Remedy: Replace in kind.

Figure 8.6 - Photomicrograph

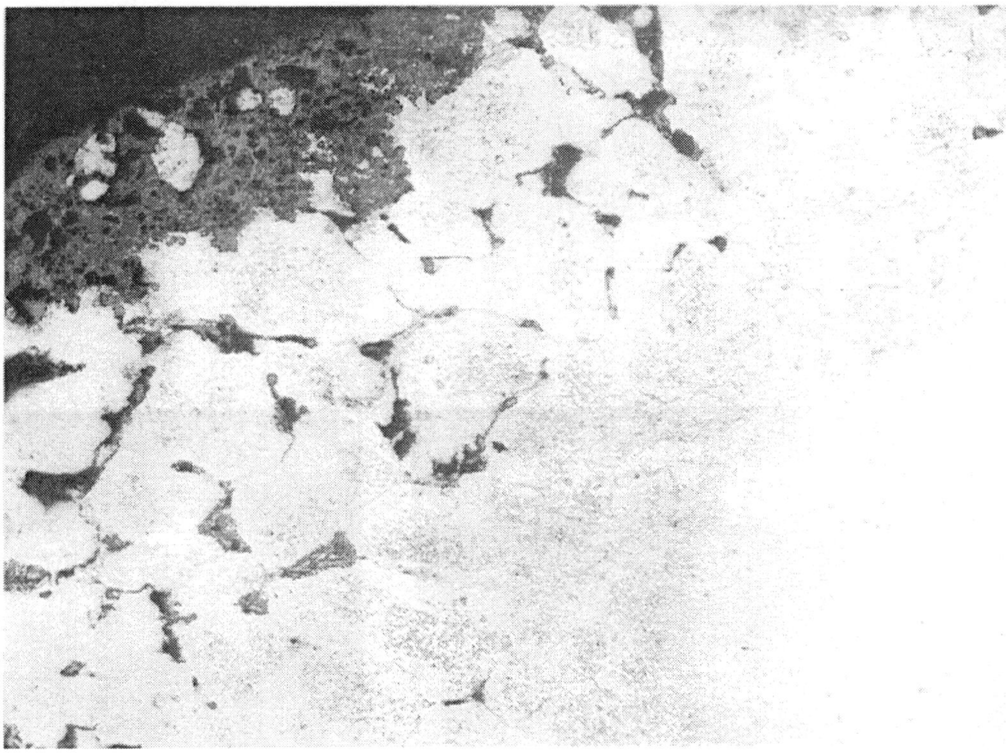

CASE HISTORY 8.7

Phenomenon: carburization
Group/ Type: SS, 35Ni-19Cr-Cb austenitic
Equipment: woven-wire furnace belt
Environment: carbon regeneration furnace
Conditions:
(a) major species: carbon particles, water
(b) minor species: sulfur, oxygen, chlorides
(c) temperature: 350 to 1,850°F (177 to 1,010°C)
(d) pressure: atmospheric
(e) aeration: 12% O_2 in hot cycle

Time to failure: 276 hours
Illustration: photomicrograph
Comments: This furnace reactivates carbon filter particles by burning accumulated organic matter out of their pores. The high temperatures lead to carbon diffusion into the wire of the woven furnace belt. As the photomicrograph shows, the carbides have precipitated densely within the grains and formed an almost continuous network at the grain boundaries. Such heavy carbide precipitation robs the alloy of the ductility required to survive the flexing that a furnace belt undergoes.
Remedy: Replace with an alloy containing higher Si and Al contents to improve the protective oxide film.

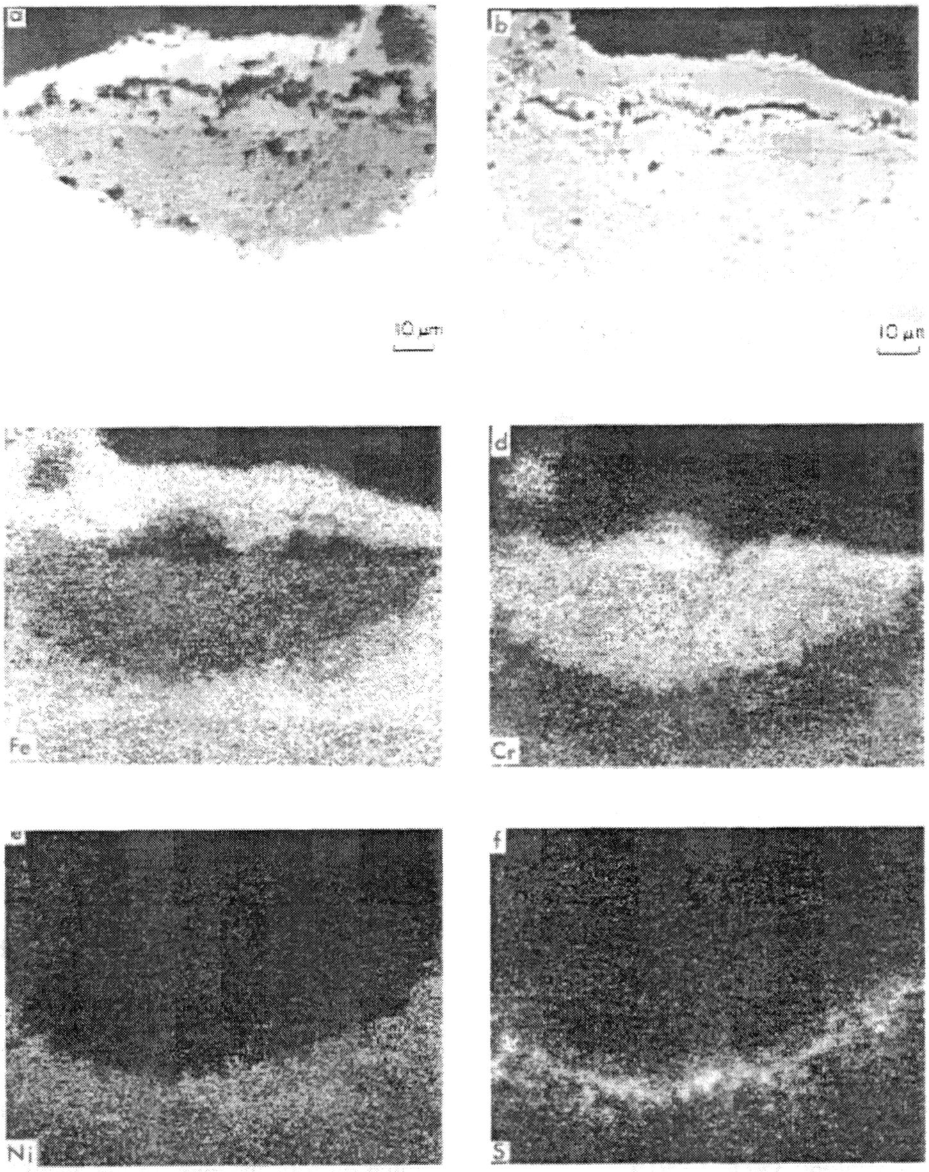

CASE HISTORY 8.8

Phenomenon: sulfidation
Group/ Type: SS AISI 310 (UNS S31000)
Equipment: combustion vessel
Environment: cyclic exposure, fluidized-bed coal combustion gas

Conditions:

(a) major species: high O_2 cycle: 95% CO_2, 5% air; low O_2 cycle: 89% CO_2, 4% CO, 6% argon
(b) minor species: high O_2 cycle: 10 to 2 atm O_2; low O_2 cycle: 0.06% SO_2
(c) temperature: 1,652°F (900°C)
(d) pressure: atmospheric
(e) aeration: as given

Time to failure: 8 hours (16 cycles)

Illustration: micro, electron microprobe element maps

Comment: AISI 310 SS showed good resistance to the cyclic conditions in the fluidized bed reactor when exposed to a high O_2 cycle first. However, when exposed to a low O_2, sulfur-bearing gas, rapid sulfidation attack occurred. A cross-section of a pit is shown in the figure. The rest of the figures show K-alpha x-ray maps of iron, chromium, nickel, and sulfur. As the maps show, the outermost oxide layers are of iron. Inner oxide layers are a mixed FeCr oxide presumed to be $FeCr_2O_4$. At the base of the pits are intergranular, chromium-rich sulfide precipitates. Nickel has been selectively attacked.

Remedy: Condition the vessel with high O_2 cycles before service in sulfur-bearing gas.

CASE HISTORY 8.9

Phenomenon: high-temperature carburization
Group/ Type: heat-resistant alloy, ACI grade HL (30°Cr, 20 Ni) (UNS J94604)
Equipment: reformer tubes
Environment: ethylene production
Conditions:

(a) major species: naptha, ethylene
(b) minor species: carbon
(c) temperature: 2,012°F (1,100°C) steady state
(d) pressure: not given
(e) aeration: oxygen free

Time to failure: 20 days
Illustration: macro, micro
Comment: Decoking cycles in ethylene furnaces must be carefully controlled with regard to both temperature and time; otherwise, furnace tube damage will result. Too rapid decoking of this furnace led to complete carburization of the tube wall. Subsequent thermal stresses produced a completely brittle fracture.
Remedy: Control temperature rise rate closely during decoking

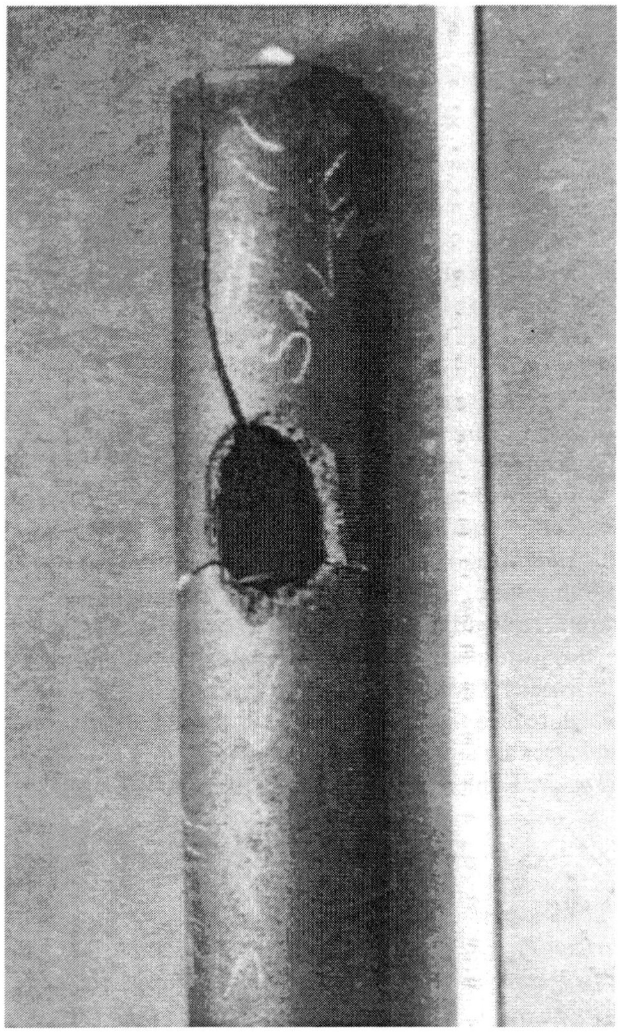

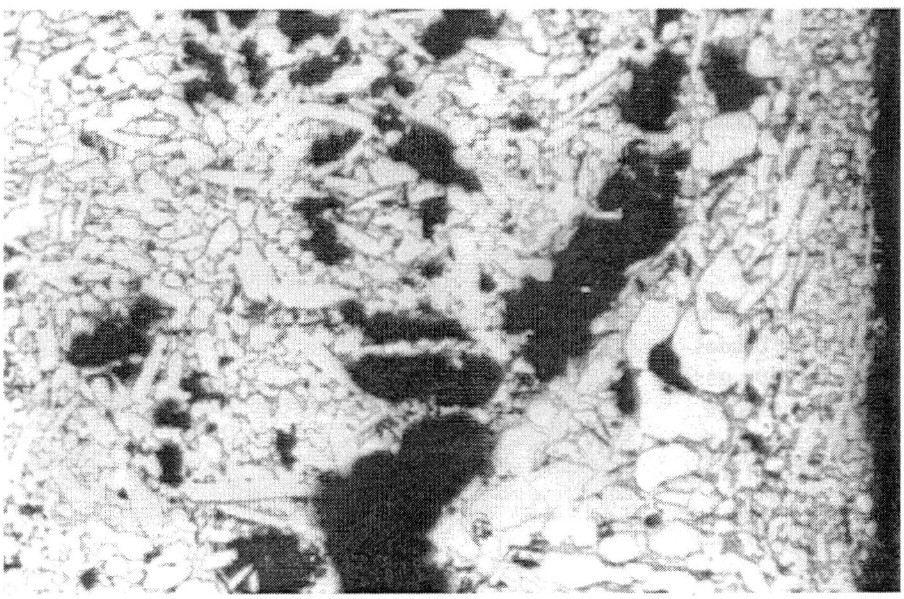

Figure 8.9 (a) -(above) Macro

Figure 8.9(b) - (left) Mixed acids etch, 100X.

CASE HISTORY 8.10

Phenomenon: creep-rupture
Group/ Type: ACI grade HL (UNS J94604): 50% Fe, 30% Cr, 20% Ni
Equipment: pyrolysis furnace
Environment: hydrocarbon gases
Conditions:
(a) major species: hydrocarbon gas
(b) minor species:
(c) temperature: 2,000°F (1,093°C)
(d) pressure: not given
(e) aeration: none
Time to failure: 7 months
Illustrations: macro, micro
Comments: Unusually high nitrogen levels (0.25 to 0.34%) in a lot of cast furnace tubing resulted in early failure. High nitrogen levels promote precipitation of a brittle nitride phase, which takes either a lamellar or acicular form. As carburization begins (a normal occurrence in pyrolysis furnaces), the nitrogen solubility of the matrix decreases. The nitrogen forms grain boundary voids, which reduce the creep strength of the tubing. Bulging and cracking signal failure.
Remedy: Limit nitrogen in furnace tubing to below 0.13%.

Figure 8.10(a) - Macro

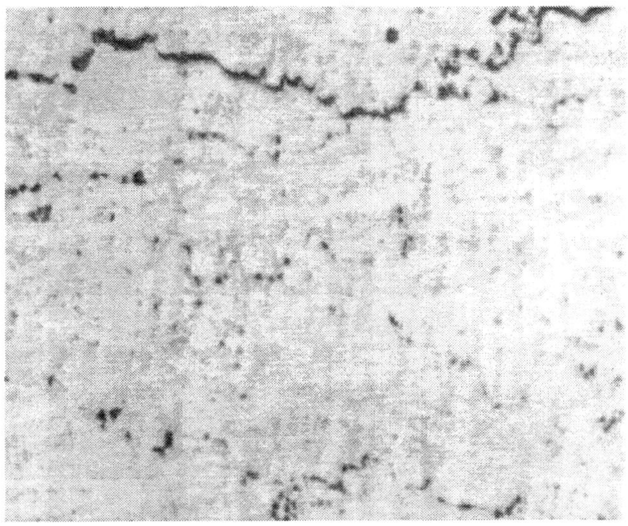

Figure 8.10(b) - Micro

CASE HISTORY 8.11

Phenomenon: hot salt corrosion
Group/ Type: AISI 316 SS (UNS S31600)
Equipment: aluminum recycling furnace recuperator
Environment: flue gases, fluxes
Conditions:
(a) major species: flue gas, chlorine
(b) minor species: NaCl, NaF, CaF_2, AlF_3, Na_2SO_4
(c) temperature: 1,400 to 1,500°F (760 to 816°C)
(d) pressure: not given
(e) aeration: none
Time to failure: 1 year
Illustration: macro
Comment: Chlorine gas is used to degas aluminum scrap, and complex salt fluxes are used to strip the oxides off the molten aluminum. Consequently, the flue gas from such an operation can be extremely corrosive. These SS AISI 316 (UNS S31600), 0.125-in. thick recuperator tubes were perforated in one year in this service.
Remedy: Replace with high temperature alloys 18B and X.
Reference: G.Y. Lai, *J. Met.*, July 1985, pp. 14-19.

Figure 8.11 - Macro

CASE HISTORY 8.12

Phenomenon: sulfidation
Group/ Type: nickel-based alloy, Inconel 600 (UNS N06600)
Equipment: furnace belt
Environment: carbon regeneration furnace
Conditions:
(a) major species: carbon particles, water
(b) minor species: sulfur, oxygen, chlorides
(c) temperature: 350 to 1,850°F (177 to 1,010°C)
(d) pressure: atmospheric
(e) aeration: 12% O_2 in hot cycle
Time to failure: 276 hours
Illustration: micro, EDX spectra
Comments: Activated carbon filter particles are periodically cycled through a furnace to burn accumulated organic matter out of their pores. Sulfur compounds in the process break down in the furnace, resulting in sulfur diffusion into the woven wire furnace belt. Grain boundary precipitates form. EDX analysis (see figure below) indicates that these precipitate particles are mixed CrNi sulfides. The sulfides, thermodynamically less stable than oxides in this atmosphere, are converted to oxides within

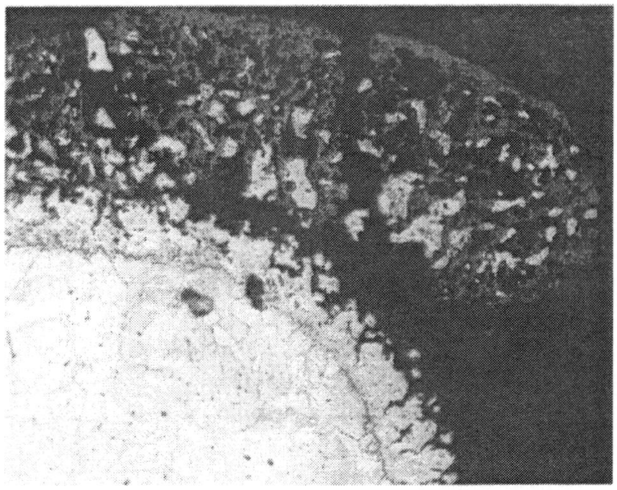

the wire. Elemental sulfur is released on conversion; the sulfur diffuses farther into the alloy to form more sulfide precipitates. Meanwhile, the oxides, brittle and weak, crack and spall off as the belt flexes around bends. Thus, a continuing cycle of degradation occurs until the wire is consumed.
Remedy: Replace with an Si-or Al-containing alloy such as AISI 314 SS (UNS S31400), UNS N06625 or UNS N06601.

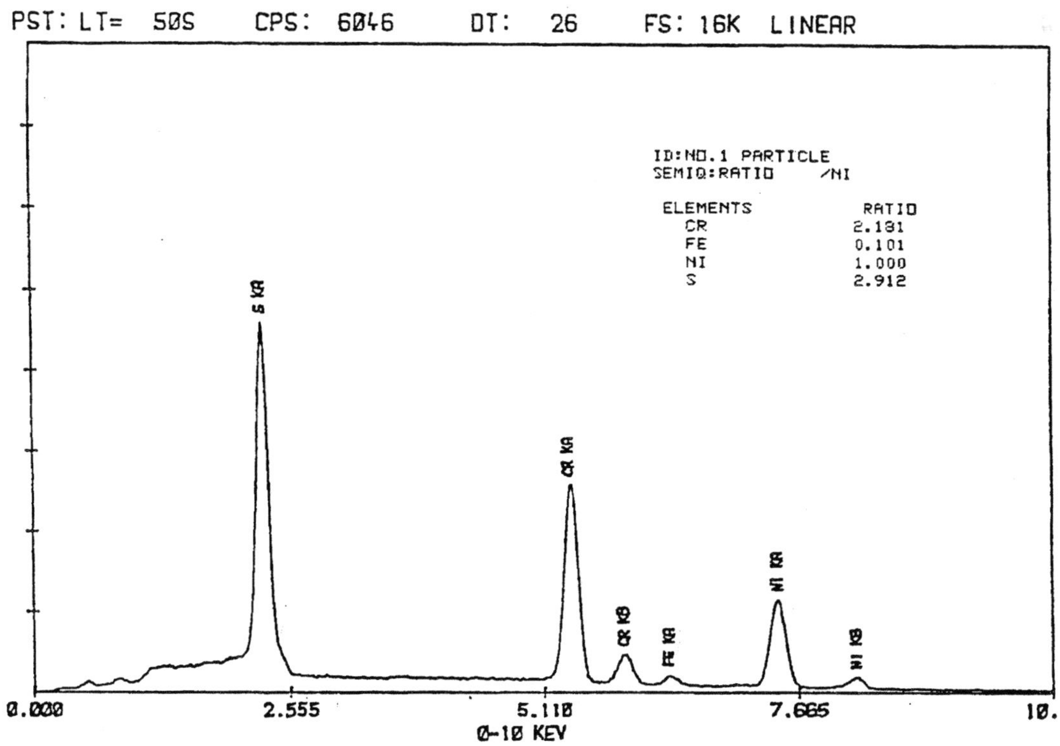

CASE HISTORY 8.13

Phenomenon: creep-rupture
Group/ Type: copper alloy, 70-30 cupro-nickel (UNS C71500)
Equipment: support grate
Environment: phosphoric acid
Conditions:
(a) major species: H_3PO_4, C_2H_4
(b) minor species:
(c) temperature: 900°F (482°C)
(d) pressure: 1,100 psi
(e) aeration: anaerobic
Time to failure: 11 years
Illustration: SEM fractograph
Comment: A 70-30 CuNi support grate suffered intergranular cracking in a highly stressed fillet area. Creep-rupture was diagnosed as the cracking mode.
Remedy: Repair weld with 70-30 CuNi rod.

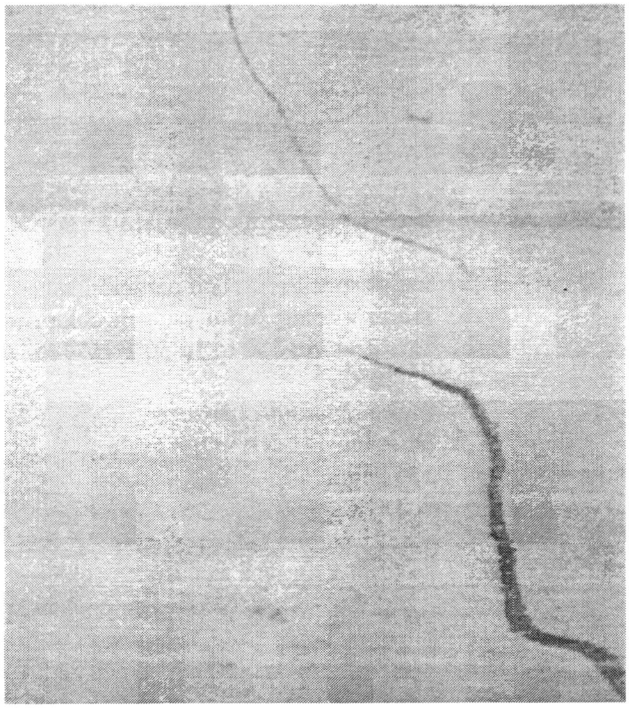

Figure 8.13 - SEM fractograph

CASE HISTORY 8.14

Phenomenon: carburization
Group/ Type: nickel-based alloys/alloy 625 (UNS N06625)
Equipment: woven-wire furnace belt
Environment: carbon regeneration furnace
Conditions:
(a) major species: carbon particles, water
(b) minor species: sulfur, oxygen, chlorides
(c) temperatures: 350 to 1,850°F (177 to 1,010°C)
(d) pressure: atmospheric
(e) aeration: 12% O_2 in hot cycle
Time to failure: no failure, 188 service hours
Illustration: SEM micrograph of polished cross-section
Comments: Activated carbon filter particles are regenerated by periodic cycling through a furnace to burn out organic material clogging their pores. The illustration shows the early stages of carburization beginning in nickel-based alloy 625. A compact, protective chromium oxide scale has formed on the surface. Beneath the scale, niobium carbide grain-boundary precipitates are beginning to form. The niobium carbide formation ties up the carbon diffusing into the alloy and allows the chromium to remain in the matrix since NbC is thermodynamically more stable than $Cr_{23}C_6$. Therefore, the chromium is still available for the maintenance of the protective oxide film.

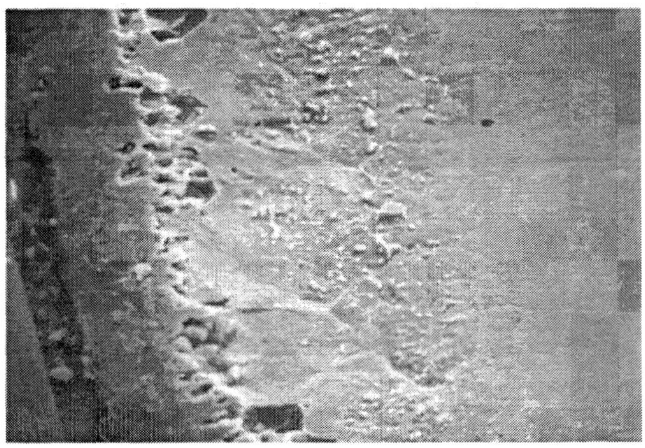

Remedy: No replacement required. However, less-expensive iron-based alloys with silicon, aluminum, and niobium additions eventually were installed to lower the installed cost of the system.

CASE HISTORY 8.15

Phenomenon: sulfidation
Group/ Type: nickel-based alloy 600 (UNS N06600)
Equipment: mineral calciner
Environment: high-temperature ore
Conditions:
(a) major species: calcium
(b) minor species: sulfur
(c) temperature: 1,900°F (1,038°C)
(d) pressure: not given
(e) aeration: not given
Time to failure: 1 month
Illustration: macro, micro
Comment: This 18-in. diam by 0.75-in. wall thickness by 21 ft long gas-fired mineral calciner ruptured after only one month in service. The feedstocks were reportedly a source of sulfur. Extensive formation of chromium and nickel-rich sulfides was observed on SEM examination of the fracture profile.
Proposed remedy: Switch to FeCrCoNi alloy 556.
Reference: G.Y. Lai, *J. Met.*, July 1985: pp. 14-19.

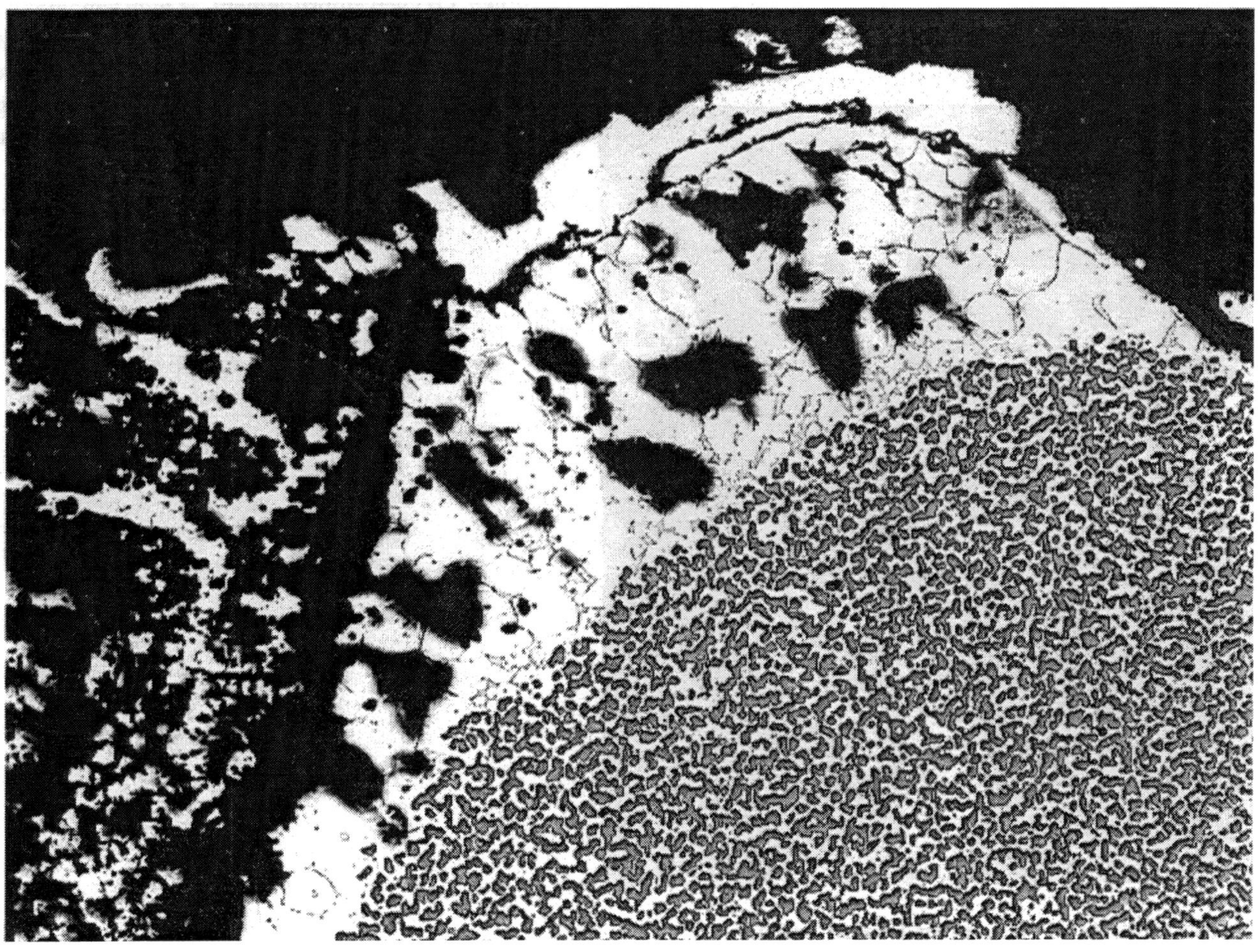

Figure 8.16(a) - Cleavage 2,000X.

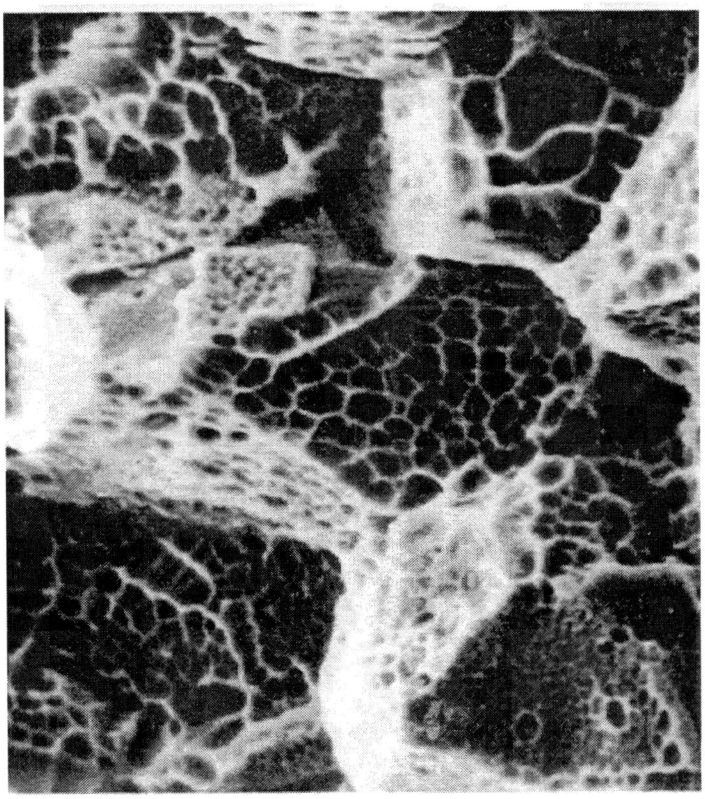

Figure 8.16(b) - O_2 embrittlement 2,000X.

CASE HISTORY 8.16

Phenomenon: oxygen embrittlement
Group/ Type: columbium alloy FS-85
Equipment: laboratory test specimen
Environment: simulated spacecraft reentry
Conditions:

(a) major species: air, water
(b) minor species: NaCl
(c) temperature: cyclic, from ambient to 3,000°F (1,649°C)
(d) pressure: cyclic, from atmospheric to vacuum
(e) aeration: cyclic, from none at high vacuum to saturated during salt bath

Time to failure: not given
Illustrations: SEM fractographs
Comment: This test specimen was alternately exposed to a high temperature in near-vacuum, followed by salt water immersion to simulate spacecraft atmospheric reentry and ocean splashdown. Oxygen embrittlement at high temperatures caused cracking by an unusual duplex mode with ductile dimples occurring cover intergranular facets (Figure). Hot-salt stress cracking by cleavage also occurred on this specimen. Final fracture was by dimpled ductile rupture.
Remedy: Insulate with refractory.

Index

F

G

K

L

O

P

R

S

Z